THE
RUT

THE
RUT

The Spectacular Fall Ritual
of North American
Horned and Antlered Animals

By Ron Spomer

WILLOW CREEK PRESS

Minocqua, Wisconsin

Photography by: Ron Spomer: Pp. 5, 6, 9, 14, 15, 16, 19, 20, 24, 25, 26, 27, 28, 30,
                   31, 32, 35, 39, 40-41, 43, 44, 48, 52, 54, 55, 56, 59, 61, 66, 68-69,
                   70, 71, 80, 85, 89, 96, 99, 102, 104, 105, 106, 112, 113, 114, 116,
                   123, 136, 143, 145.
                   Michael H. Francis: Pp. 12, 33, 46, 64, 67, 72, 77, 78, 82, 84, 87, 88,
                   90, 92, 101, 109, 110-111, 117, 118, 121, 126, 128, 130, 132,
                   134-135, 137, 144.
                   Lon E. Lauber: Pp. 94, 146, 148-149, 151, 153.
                   Erwin & Peggy Bauer: Pp. 2, 22, 36, 38, 42-43, 45A, 45B, 76, 83, 86,
                   100, 138, 139.
                   Dale C. Spartas: Pp. 75.
                   Bill Silliker Jr.: Pp. 45C, 49, 50, 120, 124-125.

ISBN: 1-57223-050-9

Published by   Willow Creek Press
               P.O. Box 147
               Minocqua, WI 54548

Designed by Steve Heiting

For information on other Willow Creek titles,
write or call 1-800-850-9453.

Printed in Canada.

Library of Congress Cataloging-in-Publication Data

Spomer, Ron.
    The rut : the spectacular fall ritual of North American horned and
antlered animals / by Ron Spomer.
    ISBN 1-57223-050-9 (alk. paper)
    1. Ruminants--North America--Reproduction. 2. Ruminants-
-Behavior--North America. 3. Sexual behavior in animals.
4. Estrus. I. Title.
QL737.U5S66   1996   96-42404
599.73'50456--dc20   CIP

Table of Contents

Introduction

Trite though it may sound, sex is beautiful. Because of sex, flowers are fragrant and delicate, birds colorful and melodious, and deer heavily antlered. Yes, we have sex to thank for the magnificent antlers and horns we admire on our wild deer and sheep. Because of sex, deer labor under the bulk of huge racks that are carefully nurtured and protected during a six-month growing period, only to be cast off a few months later. Elk actually rob minerals from their skeletons to fuel production of their annual antlers. Bighorn sheep tolerate 25 pounds of curled, hardened keratin atop their skulls 365 days a year so they can breed for a mere 14 of those days.

Thanks to sex we thrill to the whistle of bull elk in September, laugh at lovesick bull moose courting dairy cows in October, puzzle over whitetails thrashing branches in November, and watch in awe as rams slam into one another with the combined force of 2,400 foot-pounds.

Sex is such an overwhelming priority in nature that bucks and bulls will go for weeks without eating during the rut but won't miss a day of rutting to eat. A wary whitetail buck will hide in dense cover until nearly dark to avoid predators for 11 months of the year, but, come the November rut, he'll wander open fields at noon while scent trailing an intoxicating doe. And while a doe or cow will guard, nurture, and defend her young throughout the year, when in estrus she'll ignore them or even chase them away.

Without doubt, procreation is the driving force in nature, yet most of us are mystified by the varied adaptations horned and antlered beasts have made to facilitate that annual phenomenon. Even though mankind has been in close contact with deer, sheep, and cattle, both wild and domestic, for thousands of years, only in the past five decades have we applied scientific investigation in an attempt to understand their behavior, to answer the mysteries of their lives. Some of the answers are simple, some simply amazing. Others merely raise additional questions. Why antlers in some species, horns in others? Why do elk vocalize their passion while mountain goats remain silent? Why do whitetails scent-scrape and mule deer horn branches?

Why do bucks and bulls associate with females only during the short breeding period? Why don't they stick around to protect the young with their antlers, horns, and superior bulk? Why do whitetail bucks sneak away for a tryst with one doe at a time while elk bulls wear themselves thin trying to service a harem? Why does nearly every female in a herd come into estrus at the same time?

Obviously, there are many questions surrounding the rutting behavior of North

THE RUT

America's *Artiodactyl Ruminants*, starting with "What the heck is an *Artiodactyl Ruminant?*" *Artiodactyla* is the order of mammals that has an even number of hooved toes. Deer, sheep, bison, and cows are in, horses are out. The Suborder *Ruminantia* includes animals that ruminate (chew their cud). Again, think cattle, sheep, deer, and the like, all animals that also happen to have antlers or horns. Among North American wildlife this grouping includes all our familiar deer — whitetails, mule deer, elk, moose, and caribou — plus our native horned bovids, which include bison, muskox, mountain goat, pronghorn, and all subspecies of wild sheep.

Plenty of books have been published describing what those animals are. This one concerns itself with why they are. Specifically, why they do the things they do during the rut, their most active, aggressive, interesting season. This book investigates how terrain, habitat, climate, predators, and other factors combine to shape each species' behaviors and strategies for perpetuating itself, and attempts to do so with simple terms and clear, illuminating metaphors.

A good scientist avoids anthropomorphizing, the projecting of human characteristics onto animals, but I am not a scientist, let alone a good one. Consequently I have used anthropomorphism throughout this book with the belief that the average reader will better relate with and understand what are, after all, motivations and behaviors common to many species, man included. Readers should not take anthropomorphic metaphors literally. Use them as points of reference. Don't assume that cow moose are buxom or even blonde just because I compare them to Mae West. Read such metaphors for insight, not scientific accuracy. Also, terms such as baby, teenager and old geezer are used to describe an animal's behavioral age as compared to humans, not its actual age. A three-year-old elk is not chronologically a teenager, but his physiological and behavioral development approximate that of a teenage human.

The bulk of the facts, figures and theories contained in this book have been taken or distilled from scientific reports and dissertations. That does not mean they are cast in stone. There are few absolutes in nature. What we understand as truth today may become the oversights of tomorrow (we once believed the Earth was flat). To the best of current knowledge, the information herein is both accurate and subject to change as humans continue to study, interpret, and try to understand the mysterious ways of our fellow travelers on planet Earth.

Foreword

The rutting activities of big game animals, which this book is all about, fascinated already the earliest of our ancestors, the cave painters of Europe's Upper Paleolithic. These Ice Age hunters depicted courtship, displays and fighting in red deer, caribou, wild cattle, ibex etc. and they did so with uncanny accuracy. They were excellent observers who keenly appreciated tiny, but relevant details. After all, these animals provided all of their sustenance and hunting was everything in a periglacial environment. The sketches range in age from 15,000 to 40,000 years; some may even be older. Yet despite the tens of thousands of years that separate us from these early hunters, their sketches on bone or the paintings on rock walls convey clearly what they saw, as well as what they appreciated. They showed red deer stags with long, many-tined antlers, the mouth open and head and neck extended in the rutting roar. Not any stag, mind you, but the finest of the species, just the type modern artists tend to choose for their paintings. Moreover, the stags have heavy neck manes, identifying them as members of the West-European subspecies of red deer. Yes, the cave painters were accurate. A rutting reindeer bull, nose lowered, nostrils flared, is trotting along apparently sniffing the ground for female scent, just as modern reindeer bulls do. We know it is a reindeer because the artist depicted the coat pattern specific to reindeer, which differs subtly, but consistently, from that of caribou. Moreover, the kink in the antler beam above the rear tine is prominent, as is typical of European reindeer. We know that it is a rutting stag because the belly is drawn in. That only happens in large bulls during the rut, because the big bulls then virtually quit feeding. It surprises to see such accuracy. Reindeer bulls may form compact bulls-only herds just prior to the rut. An Ice Age artist sketched such a gathering. He accurately depicted "the wall" of huge antlers. He must have been impressed by that sight. I know I was when I first saw such a gathering of bulls and stared almost in disbelief at "the wall" of antlers.

The rutting season is a brief period of high activity in the annual cycle of northern ungulates. The males had been on vacation until then. They were previously secretive, nurturing their bodies for the great event. But now they burst out into showy competition for females and courtship. It is an exciting time to be in the field to see it all before the hubbub subsides, before the males, exhausted by their supreme effort, thin and haggard and nursing dozens of wounds, retire from the field to become invisible once again. I was fascinated by it all early on as a hunter. To call in a rutting bull moose is very thrilling, to play bull elk with call and antler in hand and utterly convince a bull elk that you are a rival is even more so. However, there is a deeper satisfaction in playing detective as a scientist, deciphering what all the signals meant, how they functioned and why they evolved.

On one unforgettable day, four graduate students in zoology from the University of British Columbia climbed onto the range of California bighorn sheep in southern British Columbia to watch the rut. I was one of the four. We had also rifles along, as we hoped to bag a mule deer buck to augment our slim resources. However, it never came to hunting, even though we did see some fine mule deer bucks. We were captivated by the mountain sheep. There, in the bright sunlight on glistening snow was a herd of about a hundred, mostly females and young, held in turmoil by a large group of heavy-horned rams. It went on and on. I was puzzled by the distinct signals they performed, in particular by the long runs of the rams with head held low, nose extended. Again and again a ram would run in this posture into the gathering of rams, or after a female. What did it mean? It was not until I studied Stone's sheep two years later that I found out. It was one of the horn displays of rams, a signal performed by dominant rams toward subordinates, a way of showing off, of "rubbing in" their superior status. Occasionally, however, it would be performed by a small ram toward a large one, provided the large ram was well below the cheeky upstart and could not reach to punish him. However, by then I had more mysteries at hand, such as why mountain goat males upon appearing on the rutting ranges would crawl on their bellies to the hostile females, and why they were chased away until after lengthy persistence they were accepted by the nannies for about three weeks, did the breeding, and were then chased from the female winter ranges by the aggressive females. The mountain goats and mountain sheep were side by side on the same slopes and cliffs, but the contrast in rutting behavior couldn't be greater.

This book depicts the rutting of our North American big game. It introduces the reader not only to the action, what the rutting partners and rivals do, but also what the action is all about. It offers explanations, informing the reader in a lively, engaging fashion about the detective work done over the last three decades deciphering the meaning and purpose of rutting activities. Thus I was delighted to see the huge-antlered Irish elk, *Megaloceros,* brought in to explain the large antlers and rutting strategy of caribou. Subsequent to publishing the papers available to Ron, I investigated skeletal proportions of *Megaloceros* and compared these to similar data from many other deer and antelope. This research showed clearly that, as predicted by its huge antlers, the Irish elk was the most advanced runner ever among deer, just as the caribou is the most advanced runner among surviving deer species, followed by the elk, the most running and plains-adapted of the red deer.

There is much to read and enjoy in this volume. It is the first book to put the different North American species side by side, illustrating in clear prose free of scientific jargon how different elk behave compared to mule deer, pronghorns compared to bison, or mountain sheep compared to mountain goat. Procreation is not simple and each species does it in a different fashion. However, there is reason behind these differences, as Ron Spomer lucidly explains.

This is a unique book, and I hope you will enjoy it.

—*Valerius Geist,*
Professor Emeritus of Environmental Science,
Faculty of Environmental Design
The University of Calgary

Antlers & Horns

Them bones them bones ...

If male deer didn't have antlers, a creative science fiction writer would invent them. We'd see them emerging from a UFO in a B-movie, stuck to the cranium of some other-world creature. We'd gape, stare and shudder. Ugly!

We, however, grew up with Bambi, Bullwinkle, and probably a few real deer in our backyards, so we think of antlers as majestic, regal crowns. But stop and contemplate them and you must admit that antlers are bizarre. Naked bones sticking out of an animal's head? What's the sense of that?

Most people think antlers are fighting tools designed to defend their bearers against rapacious predators. A Disney myth. If that were true, bucks wouldn't throw away perfectly good racks each year to grow new ones, leaving themselves defenseless in the meantime. And, if antlers were designed for defense, females would have the biggest and best. Yet, while bucks and bulls lounge about all summer growing new antlers, does and cows raise the young 'uns bereft of antlered protection. Of 39 deer species worldwide, only caribou/reindeer females wear antlers, and they are tiny.

The question, then, remains. Why antlers? The answer is tied to the annual rut, and it seems so recklessly extravagant of Mother Nature that it may surprise you.

But one must first understand what antlers are and how they differ from another familiar cranial adornment — horns.

The most dramatic distinction is that horns are permanent, growing slightly larger each year, but antlers are deciduous. Milk cows have horns. Bison and muskoxen have horns. Sheep, goats, antelope, both male and female, are all horned. Only the heads of deer, the Cervidae family, are antler crowned. In North America that includes elk, moose, caribou, white-tailed deer, mule deer and all their subspecies.

Superficially, antlers are like maple leaves. They grow, die, and fall annually. Unlike leaves, antlers play no role in producing food for their hosts. They don't turn brilliant colors in October, either, but they make up for that with their size and symmetrical, branching forms. Also, antlers drop in winter or early spring instead of autumn. In March or April a fresh set, like new leaves, begins growing.

Here's how it works: Both male and female deer are born with microscopically visible pedicles lying beneath vascular periosteum tissue under the skin of their heads. This periosteum is the actual growth bud of the antler. It is activated by a small dose of testosterone when a fawn is several months old, and begins laying down calcium to transform the pedicles into bony extensions of the skull. New antlers will sprout from these bumps each spring for the rest of that animal's life, and the pedicles will enlarge in circumference annually. The antlers will grow until late summer or early fall, then die and eventually fall off in an annual cycle that corresponds to the breeding cycle of the females of the species. Females produce young each year. Males produce antlers. Hardly seems equitable, yet both are necessary, as we shall see, and antler production can be as much of a physical drain as growing a fetus.

Sheep, like this bighorn, goats, pronghorn antelope, muskox and bison all grow horns instead of antlers.

Pedicles can be felt just under the skin of male fawns and calves in late summer, and in some healthy individuals can be seen as small bumps pushing up the skin and hair. Since long before Davy Crockett and Dan'l Boone, hunters have called these "button bucks." In well-nourished males, pedicles may even grow through the skin and harden into tiny spikes by the animal's first autumn.

Naturalists have long known that if a pedicle is broken, perhaps in a fight, and a piece of it shoved to a new location on a buck's head, an antler will grow from the relocated piece as well as from what is left of the original pedicle. Accidental trauma to pedicles in the wild has led to three- and four-antlered bucks. Curious about this, researchers have snipped bits of periosteum from pedicles, grafted them onto other parts of a buck's body, and grown antlers on noses, legs, hips, even ears. It seems all periosteum needs is a blood supply and a shot of testosterone to begin laying down a pedicle.

Because pedicles increase in diameter throughout a deer's life, the older the buck or bull, the thicker its antler base and, given adequate nutrition, the larger the rest of the antler. Spindly beams are the products of youth because young animals must funnel nutrients into growing their skeletons, leaving little to spare for antlers. Whitetails and mule deer are fully mature and produce their best antlers between five and nine years of age. Elk grow increasingly larger sets each year until age 10 or 11, although poor forage can reduce any set.

Moose are the ultimate antler manufacturers. Within three to five months a mature Alaskan bull moose can grow more than 90 pounds of palmated bone stretching six feet wide. In contrast a mule deer requires five months to grow its relatively small antlers, and an elk six to eight months to produce his. As biologist Victor Van Ballenberghe noted after his study of moose in Denali National Park, "This rate of bone tissue formation by an adult animal (moose) must be without parallel in the animal kingdom."

Antlers grow three to seven times faster than body bones. Because of this rapid growth and the uniqueness of antlerogenesis (biologists' jargon for the antler growing process), scientists study it for insights into bone growth and limb regeneration in humans. Though no one yet fully understands antler production, research indicates growth is controlled by a complex relationship between *photoperiodicity* (changing daylight length), the brain, several glands and various hormones, the most significant of which seems to be testosterone. This male hormone is produced primarily in the testes but may also arise from the adrenal gland. Castrate a male deer before it begins producing testosterone and it will not grow pedicles, and probably never grow antlers. Inject the same neutered animal with testosterone and bingo! Pedicles begin to sprout and antlers to grow at nearly any season. Administer testosterone to any female deer and she will begin growing antlers.

This spike elk and other young deer grow "spindly" antlers because their bodies funnel nutrients into their growing skeletons. With maturity, their antlers will grow larger.

Hormonally imbalanced females of all deer species have been known to grow antlers, but that is not a usual circumstance. And, just because they normally grow antlers, caribou cows are not the "bearded lady" of the deer world, overdosed on male hormone. Other substances, perhaps steroid hormones from the ovaries or adrenal glands, control antler production in adult caribou cows and possibly bulls as well. Testosterone does not play exactly the same role in antler production in all species.

Although a quick spike of testosterone in early spring initiates antler growth, too much of the stuff stops it. Throughout the summer growing period testosterone levels remain low and estrogen levels fairly high. This essentially feminizes males, permitting bucks and bulls to associate in fairly companionable bachelor bands with minimal fighting. They even look like females with their thin necks and fight like females with flailing hooves. But swelling bumps atop their heads, covered with fuzzy hair called velvet, reveal their sex.

Within this velvet skin "wrapper" are nerve fibers, blood vessels, and a rapidly growing matrix of cartilage that is 80 percent protein and a variety of trace minerals including calcium, phosphorus, iron, manganese, silicon, bismuth, titanium and even gold (fortunately in concentrations too low to support commercial exploitation). Nerve bundles extend to all those fuzzy velvet hairs near the growing antler tips — some 6,000 per square inch — and allow the buck to feel obstructions and thus avoid banging his delicate antlers against them. This sensitivity enables a buck to learn and remember his antler size so that, even after they've hardened and the nerves have atrophied, he can precisely scratch an itch on his rump or twist his head to pass

between branches without touching them.

Because of an antler's unsurpassed growth rate, it must be fueled with a rich blood supply delivered through arterioles branching from the supraorbital artery over the deer's eyes. These divide into a dozen temporary arteries that ascend each growing antler just under the velvet skin, then branch into the heart of the growing beam. Blood flow is so great that velvet antlers are hot to the touch, registering nearly 104 degrees on their surface. The main arteries can constrict rapidly to stop hemorrhaging in less than a minute.

Amazingly, growing antlers, lengthening as much as 1/4-inch per day in mule deer and 3/4-inch per day in moose, require more calcium and phosphorus than their host can supply. Studies have proven that the healthiest bull elk eating the highest quality forage often cannot ingest enough to supply the need. Yet the antlers grow. How? Through cyclical osteoporosis. Deer literally rob their body skeletons to grow antlers that they'll abandon a few months later. During peak antler production, resorption of rib bone minerals is nearly 25 percent. This drops to less than three percent near the end of the growth period. A big bull elk may have to pull six pounds from his skeleton to supply his growing rack. After the antlers harden, borrowed skeletal minerals are quickly replaced, and by autumn the prolifigate stag is at his physical peak — strong, fat, drunk on testosterone and full of fight.

Essentially bucks and bulls are slaves to their antlers. In order to grow them, they must eat young, tender shrubbery and forbs

The healthiest bull elk, eating the best quality forage, still cannot provide enough nutrients to adequately fuel his antler growth. Therefore, the elk's body robs its own skeleton of minerals.

(herbaceous, non-woody plants that are not grasses, i.e. wildflowers, weeds, clover), which are higher in calcium and phosphorus than are grasses. To find these plants bulls wander in small bachelor bands high up mountains, avoiding competition from herds of cows and calves, but simultaneously losing the herd's protection against surprise attack from predators. One can predict whether a ruminant species will sport antlers or horns based on where it lives and what it eats. Grazers live in the open and wear horns. Browsers live in brush or forests and wear antlers. Caribou might appear to be an exception, but the tundra is covered with dwarf birch, willow, blueberries and lichens.

Regardless the species, if a mature deer produces a short, narrow rack with few tines, he either didn't eat well during the critical growing season or, rarely, he inherited poor genetics. Genetically weak antlers tend to be self limiting. Small-antlered bucks do less breeding than large antlered bucks, thus there is a continuing trend toward larger antlers.

Nutrition is by far the most critical factor in antler growth. The better the summer forage, the bigger the antlers. Males past their prime usually regress and produce weak, few-tined antlers, but their main beams remain heavy because they sprout from large pedicles. Most bucks contribute their protein to predators, though not willingly, before they age enough to grow regressed antlers.

Manufacturing antlers and competing for breeding rights is exhausting work that makes a stag old before his time. Males rarely live more than 13 years. Females have been known to double that span.

After the summer solstice, a buck's pineal gland detects decreasing day length and signals the testes to begin producing more testosterone. Rising levels of this potent chemical eventually trigger antler hardening (death), velvet stripping, brush thrashing, and the aggressive behavior seen during rut. After the breeding season, testosterone levels fall and so do the antlers. Hardening progresses quickly from the bottom up. Cartilage is mineralized, then ossifies, becoming true bone. Bucks injected with extra testosterone in mid summer will strip velvet and begin hostilities before ossification is complete and while blood is still flowing freely into their antlers.

By manipulating the ratio of daylight to dark, researchers have tricked deer into growing and casting as many as four sets of antlers in one calendar year. When elk from the northern hemisphere were transplanted to New Zealand south of the equator, where winter and summer seasons are reversed, they quickly switched their antler growing and casting cycle to match the new seasons. Small whitetail subspecies living near the equator where day length does not vary seasonally and there is no fixed breeding season (females may mate in any month) do progress through an annual antler growing/casting cycle, but on individual schedules depending on when they were born. When kept in northern zoos, these same deer retain their individual cycles rather than respond to the photoperiod as do native northern deer. Some sort of internal biological clock is setting their schedules.

If northern whitetails are placed under an artificially-controlled lighting regimen of 12 hours light, 12 hours dark, the results are even more curious. If they enter the experiment in autumn with hard antlers, they'll keep those bones for as many as four years if the light/dark regimen is maintained. This suggests that decreasing daylight is required to stimulate antler casting, but nature is seldom absolutely predictable. Conduct this same experiment beginning after the winter solstice (December 21), and bucks shed and regrow their antlers the next spring right on schedule. But they stubbornly hang onto those new antlers for years if the unnatural lighting continues. On a schedule of eight hours light/16 hours dark or vice versa, or even constant light, bucks persist in their normal 10-month antler generating/casting schedules, but at irregular intervals. One buck might start growing his antlers in August and drop them in June, another in January and drop in November, the very month wild whitetails use them the most. This is an example of an internal clock or circannual rhythm not fully understood by science. Fortunately, wild deer don't have to worry about any of this because the Earth and Sun keep regular hours.

Now that we have some knowledge of the process and effort required to produce antlers, we can better appreciate their role as "luxury organs," so called because they aren't necessary for survival. They are, however, required for procreation. Without antlers, male deer find it almost impossible to successfully woo females.

Antlers are roughly analogous to sports cars and gold chains among human males; mostly useless, but great attention getters. The difference is that the ostentatious trappings of wealth can be ill-gotten by humans (inheritance, drug dealing), so they don't accurately reflect the genetic potential of the displayer. Deer, however, must earn their antlers, so they accurately signal a buck's genetic potential. A head stacked with bones proclaims: "Here is a super forager metabolically efficient at converting nutrients into flesh. Here is an old survivor who has avoided predators, viruses, automobiles and walking off cliffs. Here is the stud to father your fawns." In short, antlers are sex symbols.

Horns, on the other hand, are more utilitarian. Consisting of keratin, the same protein in hair and hooves, horns grow throughout an animal's life from the bottom up, increasing in girth and length simultaneously. The horn core is an extension of the skull plate, genuine bone usually heavily pneumated or honeycombed with air pockets and struts. These provide strength and absorb shock. The core is laced with a vascular network that fuels what can be prodigious summer growth of the external keratin layer of the horn called the sheath. Some species of sheep manufacture up to four pounds of horn in a single summer. Pull this outer sheath from the core (possible only when the animal is dead) and it becomes the famous hollow horn used by man over the millennia to blow alarms, scoop gravy, and store gunpowder. In winter keratin production declines dramatically, resulting in external constrictions or annual rings that, like the rings of a tree stump, indicate age.

Some horned animals are able to dissipate as much as 12 percent of their body heat through their horns, which has lead to speculation that horns evolved as thermoregulatory devices. Behaviorists reject this idea, noting that females do not grow horns as large as those of males, yet live in the same climates. Also, some of the largest horned animals live in the north where there is more need to conserve body heat than unload it.

For whatever reason they evolved,

horns serve several purposes. In some species such as the African eland they are tools to knock down overhead foliage. In others they are short, sharp stabbing instru-

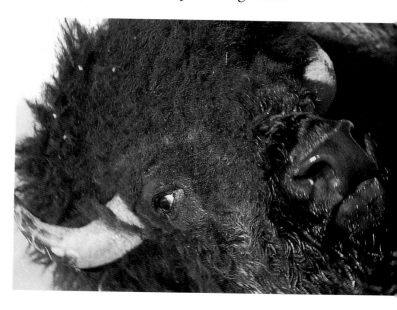

Horns have many uses. Bison use theirs to lock heads in shoving matches to determine dominance.

ments primarily for offensive attack or defense against predators. Bison use their broad, hooked horns to lock heads for shoving matches to determine dominance. And sheep, the most extravagant horn growers of all, have evolved massive horns that parallel antlers in function.

The advantage of horns over antlers is that they do not have to be regrown each year. The disadvantage is that they cannot be repaired if broken. Fortunately, most horns are massive and sturdy enough that they rarely break, which is important, for horns, like antlers, play a significant role in the mating behavior of North American ruminants.

Pronghorn

Born to run

The pronghorn antelope is famous for three things: peculiar horns, incredible speed, and the wrong name. Despite its common moniker, North America's only antelope isn't an antelope at all. Rather, it is a family unto itself, *Antilocapridae*, literally antelope-goat, because it blends features of both families. Several *Antilocapridae* ancestors roamed the North American plains millions of years ago, some sporting four horns, but all gave up the evolutionary battle, leaving today's pronghorn the last of the line.

This critter's odd horns are one of the anatomical features that excuse it from the antelope tribe. Horns are not supposed to branch or fall off every year, but no one has told the pronghorn. Granted, the forward-projecting prong on a pronghorn's horn isn't much of a branch. Compared to the tines of a mule deer or an elk they are mere bumps. Nevertheless, they extend at right angles several inches beyond the center of the main horn, and no other horn in the animal kingdom does that. African antelope horns corkscrew, spiral, twist, and turn into wonderful shapes, but none branch. Nor do any fall off every year. Actually, neither do the pronghorn's. At least not entirely. After the rut in late October or early November the outer sheath slips off, but its bony core remains firmly attached as a permanent extension of the animal's skull. From the tip of this core, covered with black cartilage, a new horn sheath immediately begins growing both upward and downward simultaneously. Like all horns, it is made of keratin, not bone.

Speed is the pronghorn's claim to fame. At bursts of 60 miles per hour and sustained 40-mph cross-country running, this is the fastest mammal in the world (discounting the short-range sprint of the cheetah). A prime buck standing on tiptoes might measure three feet at the shoulders and weigh 125 pounds after a full meal with mud on its feet, yet its skinny leg bones withstand more crushing pressure than an 800-pound steer's. Its lung capacity is three times greater than a domestic goat's. It's heart and trachea are twice the size of a 170-pound Olympian marathoner's. All of this physiology is designed to push fuel (oxygen) to the pronghorn's engine (muscles), which are unusually rich in mitochondria, the cell structures that chemically burn oxygen. This animal is built for speed, and, believe it or not, its sex life has much to do with that.

Ever since man first laid eyes on the

THE RUT

pronghorn, he has interpreted its reproductive shenanigans anthropomorphically, chauvinistically, and incorrectly, starting with William Clark who wrote the following in his journal after his partner, Meriwether Lewis, unsuccessfully stalked a band of antelope along the Missouri River in present day South Dakota:

"After many unsuccessful attempts, Captain Lewis at last, by winding around the ridges, approached a party of seven which were on an eminence toward which the wind was unfortunately blowing. The only male of the party frequently encircled the summit of the hill, as if to announce any danger to the females, which formed a group at the top."

Given the social fashion of 1804, we can excuse Mr. Clark for assuming the male pronghorn stood guard over the

During the rut, pronghorn bucks display, herd, chase, defend and maintain as tight a hold as possible on their harem.

more defenseless "ladies." We have since learned, of course, that buck pronghorns are not their sisters' keepers. What Lewis saw was not altruism, but stereotypical herding behavior, a purely selfish act. The only thing Meriwether's buck was announcing was his intention to keep those frisky gals rounded up and out of the amorous reach of other males. The only thing he was guarding was his chance for romance and a place in genetic history.

A pronghorn buck's urge to control does during the rut is so overpowering that he'll literally run himself to death trying. Researchers have documented bucks lying down to rest just four minutes during 36 hours of observation in the peak of the late-September rut. They don't waste much time feeding at this season either. Fully 93 percent of their days are spent displaying, herding, chasing, defending, and otherwise maintain-

ing as tight a hold as possible on their bevy of does. As a result, they enter winter with practically no fat reserves and often die from malnutrition, which is no great loss to the rest of the herd. Nature is efficient, not kind. After a buck's seeds are planted, he's expendable. In fact, one less mouth to feed means better survival odds for the pregnant females and their growing fetuses.

Noting the exhausting lengths to which bucks push themselves in this harem-holding effort, hundreds of cowboys, hunters, and naturalists over the centuries developed an obvious but inaccurate explanation of the species' mating behavior. To this day it is widely believed that the biggest and meanest buck in the neighborhood rounds up as many does as he thinks he can handle, herds them mercilessly, and drives all competing males away until his breeding chores are completed for the year. To the victor the spoils.

Sorry, guys. Despite a buck's most heroic efforts, it is the female that decides who mates with whom. But give the males credit for trying. They pull every trick in the book including territorial defense, harem defense, and isolated courtship. Still, it's ladies' choice.

According to recent research, pronghorns appear to practice three or four courting strategies: territoriality, harem-holding, a combination of those two, and one-on-one quiet courting. The territorial style seems to function best in stable populations with an even sex ratio and a broad range of age groups. Here's how it works: An old, mature buck starts plotting his late-summer liaisons as early as March. This is when he begins the arduous process of carving himself a territory

of one or two square miles — some say as large as five square miles in marginal habitat — on the bleak plains. Even at one square mile, this is a huge slab of property for a single animal to guard, which may explain why only mature bucks older than four normally do it. Perhaps only they have developed the skills, knowledge, and physical endurance to pull it off.

No one knows exactly how a buck chooses which chunk of country to defend. Odds are it's based on long experience. If he has survived four or more years, he'll probably know where the best combination of forage and water is, where large groups of does congregate during the rut, and where the lay of the land permits easy defense. Often a range of hills, a creek, or a deep ravine will outline a side of his spread. Most likely he inherited the territory from a previous buck who in turn assumed possession after the demise of an even earlier occupant. Like real estate, the key is location, location, location. However, when a grim winter or intense predation reap high mortality among older males, traditional territorial boundaries may fluctuate widely or break down altogether as inexperienced youngsters fumble for control.

Once a buck has selected his territory, he begins patrolling and marking its boundaries, much as a human would set corner posts and string a fence, for he is surrounded by bachelor bands of younger males who live in limbo between master buck territories or in odd corners of marginal habitat that no one else wants. Instead of steel posts and wire, a dominant buck sets a line of scent posts and scrapes, reinforcing them with ritualized behavior and displays readily understood by all on the outside looking in. These are nature's No Trespassing signs, and it is assumed they have evolved to keep other bucks at bay and minimize hazardous fighting.

Among a pronghorn buck's most common territorial behaviors is subauricular marking, an olfactory defense. The prominent, black patch of hair directly below the buck's eye and behind his jaw — what most folks call a cheek patch — covers the subauricular gland, which secretes a pungent oil detectable even to humans at considerable distance. It's what gives the pronghorn its distinctive "stink." Territorial bucks smear it generously on weed stems, branches, flower stalks, even grass clumps throughout their territories and along territorial boundaries. We assume that other bucks

A pronghorn buck patrols his territory, similar to how a human would set corner posts and string a fence.

smell it and back off, but no one knows for sure. Does are often seen sniffing vegetation marked by a buck's subauricular gland. Perhaps the scent attracts them as much as it repels males. Maybe it reassures females that they are within a master buck's territory and will be amply protected from overly aggressive courtship by inexperienced young bucks. Only pronghorns know for sure, and they're not talking.

Another gland that probably does have something to do with soothing or sexually stimulating females is the caudal gland, located on the buck's back, just forward of the white rump patch. He erects the hairs covering this gland during his courtship approach to a doe, giving his coat a distinct cowlick. Again, no human knows exactly what this gland does.

It is difficult for humans to interpret olfactory communication among ani-

A buck smears brush with oil from its subauricular gland, which helps mark territorial boundaries.

mals. Our own scenting abilities are so poorly developed that we can't even detect many odors, let alone discriminate among them, gauge their intensities, or guess their nuances. A scent barely perceptible to us might be absolutely intoxicating to an antelope. Or frightening. Even if we could smell every pheromone wafting from a pronghorn, we wouldn't understand it. We can't even decide what our own body odors are supposed to signify, preferring instead to mask them with perfumes and deodorants. Researchers must study animal responses to scent marking intently and repeatedly to form valid interpretations, and even then they're often guessing. Thus far, it appears that subauricular marking by pronghorns primarily defines a buck's territory.

The pronghorn's second territorial signpost may be both a scent and a visual marker. The buck creates it by first sniffing the ground. Exactly what odor he detects has not been determined. The scent of another buck? Doe urine? Regardless, after sniffing he paws the spot with a front hoof, removing vegetation from a square foot or less of ground. He steps over the scraped area and urinates into it, then squats and defecates into it. Biologists call this SPUD (sniff, paw, urinate, defecate). The result is a fairly obvious visual marker, at least to humans, probably to other pronghorns, and the scent value is also assumed to be significant. Biologists postulate that interdigital glands between the buck's hooves may deposit additional identifying odors during the pawing process.

These boundary markers don't seem to immediately frighten competing males away, but they give them an initial warning: *You Are Entering Private Property. Keep Out. This Means You.* Or something like that. To reinforce these markers, the boss buck mounts regular patrols, "showing the colors" throughout his territory, often stopping for long periods and staring out across the plains, head high, king of his domain. He keeps a sharp eye for trespassers, and early on there are plenty of them as other bucks search for their own territories. "The price of property is eternal vigilance," might be the pronghorn buck's motto.

When a competing buck enters another's territory, he is met with a hard stare. Bucks have been seen to stand for nearly a half hour staring down an opponent from a considerable distance. Sometimes this works, mostly on young males who turn and walk away rather than face the consequences. When it doesn't, the property holder resorts to an auditory threat, the snort-wheeze, a rapid series of a half-dozen explosive snorts that can be heard across a mile of open plains on a calm morning. If the intruder still doesn't get the message, it's time for some aggressive brush beating. The resident male will attack a sagebrush, yucca plant, or a clump of grass with his horns, twisting and thrashing in mock anger as if to say: "There but for the grace of God go you, buster. Best git while the gittin's good."

SPUD behavior practiced by bucks includes sniffing and pawing (left), urinating (below) and defecating (bottom). Facing page — A buck parades the extra adornment of grass in his horns.

Frequently clumps of grass or brush hang up on a horn, and when it does the buck seems to parade with the extra adornment, which may act to visually enlarge his horns, further intimidating the rival. Horn thrashing may also be a release of excess energy or frustration early in the rut, for bucks perform it regularly without an audience.

Thrashing and snort-wheezing are often combined with subauricular and SPUD marking. If the interloper remains, he is then subject to close approach and false charges. If he breaks and runs, he will be chased at prong

point beyond the territorial boundary, but if he is pursued into his own territory, the roles reverse and victor becomes victim. Speed among pronghorns is good for more than avoiding predators.

When two physically matched bucks dispute territory, threats and posturing escalate to violence. The pair parade side by side, heads down in a parallel walk. In this position each maximizes his body and horn size to intimidate the other. They lower heads and aim horns purposefully at each another. Finally, too full of themselves to back down, they lock horns and begin a classic wrestling and shoving match until one senses the other's superiority and flees. Interestingly, researchers have found no correlation between winning fights and breeding success, nor between horn size

Pronghorns bucks fight, but researchers have found no link between winning fights and breeding success.

and breeding success. Nevertheless, bucks keep fighting. Prolonged battles are not common, but short matches are. Bucks suffer frequent wounds as horns slip, punch, and gouge, but bucks rarely impale one another. Studies have recorded evidence of fight wounding among 30 percent of a local buck herd. That rate would undoubtedly be higher if their horns weren't so cleverly designed to minimize more serious injury.

If it were advantageous to the pronghorn species that males incapacitate or kill their competitors, their horns would be long, straight, and sharp. Instead they are curved inward at their tips, effectively blunting them as impaling weapons. Additionally, the prongs halfway down the horns block opponents' horns, preventing eye and facial injury. Lastly, a pronghorn's head and neck skin is thickest where it is most likely to take a hit during head-to-head conflict. All of this

is classic weapons evolution toward ritualized fighting designed to settle disputes and establish dominance without fatalities. Virtually all antlers are designed for ritual fighting, but not all horns. A mountain goat's horns, for instance, are lethal spears. Thus, billies do not fight head to head. Bison horns hook deep into flesh and tear gaping wounds but are intentionally short to prevent deadly impaling, thus bull bison do fight head to head. Sheep horns are the ultimate in ritualized fighting horns. They have evolved into blunt instruments for butting contests for asserting dominance through power and endurance rather than by inflicting debilitating injury. Pronghorn and sheep horns are examples of convergent evolution with deer antlers — two different organs in two different mammalian families evolving toward a common function.

All of this territorial pronghorn posturing and squabbling is fascinating and entertaining, but it would be a waste of time if it didn't give the territorial male some kind of mating advantage and improve the species' chances for long term survival. Clearly, bucks must succeed in passing on their genetic penchant for guarding a territory or nature would long ago have weeded them from the gene pool. Exactly

what are the benefits? At least two things work to the territorialist's advantage. One, if he successfully keeps competitors away from his turf before the rut, they are more likely to stay away during the rut, leaving females within his kingdom little choice in mates. This isn't 100 percent successful by any means, but it must work most of the time because the system is still in place. Two, by commandeering top-quality habitat and protecting the females within it from excessive harassment by competing males, a boss buck inadvertently provides them an environment with less energy drain, maximum energy gain, and optimum opportunity to successfully raise their fawns. A peaceful existence, quality forage and plenty of it — what more could a doe prong-

PRONGHORN RUT: FACTS ON FILE

Family:	Antilocapridae
Genus:	Antilocapra
Species:	americana
Weight:	Males 110-140 lbs., females 90-110 lbs.
Height:	32-40 inches at shoulder.
Horns:	Males 13-16 inches, females 3-5 inches.
Rut:	Late August through mid-October.
Habitat:	Shortgrass prairies, plains, mountain grasslands, deserts.
Range:	From 100th meridian (central Dakotas) west to Oregon and California, southern Saskatchewan to northern Mexico plus Baja.
Society:	Doe/fawn bands, bachelor buck bands and lone mature bucks in summer, large mixed herds in winter.

horn ask for?

Females, however, represent only half of the gene pool. What evolutionary advantage do their fawns get from a territorial buck that they wouldn't get from breeding with an interloper? The answer is health, vigor and probably intelligence. Because a territorial buck is an old buck, he has proven he can outwit predators, avoid accidents, find food and water in seasons of drought, survive winter storms, and efficiently convert forage into flesh and bone. Because he is able to define, mark, and defend his territory for six months or more, he has strength and endurance. In short, he has proven his genetic superiority.

Successful though the territorial mating system may be, pronghorns in many areas, especially where most males are less

Pronghorn bucks herd their harem like a cutting horse, preventing does from leaving but letting new members join.

than four years old, use the harem-guarding system instead. In July or August, doe/fawn family groups begin to coalesce into small herds. Bucks begin trying to control these bunches, herding them like a cutting horse to prevent any from leaving, but always letting new members join. After watching a pronghorn buck sprint, turn, cut, stop, and drive an escaping doe back to the fold, many a cowboy has wished he could saddle the little beast and ride him. By early September a mature buck might find himself with two dozen potential mates and a major problem. How can he keep them all to himself? The birth sex ratio of pronghorns is roughly even, which means there are a lot of frustrated bucks waiting in the wings. And they can raise hell with a polygamist's plans.

As the breeding season nears, a herd buck comes under assault. Feisty teenagers increasingly test his defenses.

"Inciter" does intentionally move away from the harem and invite chases and fights between the harem buck and others.

Those nearly his size often rush the harem, hoping to cut out a filly for their own. Naturally, he races to rout the interloper, often chasing him a quarter mile or more. But while he is thus engaged, other bucks slip in to try their luck. As many as a dozen satellite bucks may surround the harem and launch multiple raids. Young bucks aggressively court temporarily unguarded females, even try to mount them, but females always reject them with vigorous head shaking or by running away. A classic doe escape trick is to urinate. When the buck stops to scent-check her urine with a classic lip curl, the doe streaks out of there.

When a herd buck loses control, females lose body condition. Their resting periods can decline by nearly 50 per-cent. Even their browsing is interrupted, and that doesn't please them. So while the boys are competing and fighting like a bunch of rowdy sailors, the girls are literally stepping out, leaving the harems and asserting their ultimate control of the mating game.

During a study at the National Bison Range in Montana, researchers from the University of Idaho uncovered three amazing doe pronghorn mating strategies — sampling, inciting, and quiet rutting. Each tactic showed that, contrary to hundreds of years of popular belief, males, whether they guarded territories or harems or a combination of the two, were neither controlling the females nor deciding who would mate with them. Instead, each doe was making her own mate choice.

"Sampler" does rotated between harems every one to five days, increasing switches as their fertile period, or estrus, neared. Understandably, they would always leave a male that failed to defend a suitable tranquility zone around the harem. But, surprisingly, 71 percent of the time they left for no apparent reason, which suggests they were actually "shopping" for a suitable mate. Most returned to mate with a buck they had visited within the week before their estrus.

"Inciter" does are the vivacious teasers of the bunch. They not only sampled from harem to harem, but during their estrus would intentionally move away from their harem boss to incite chases and fights between him and bordering bucks. Inciters always watched these competitions and immediately mated with the winner. If other bucks tried to court them while the chase raged, they flatly refused them.

A courting pronghorn buck swings its head from side to side, twitching his upper lip while making fawn-like sucking sounds.

"Quiet" does really blew man's notions of pronghorn mating behavior out of the water. These does would move to an isolated area with a single male a month before their estrus and remain with him until time to breed. Researchers hypothesized that females with low energy reserves adopted this strategy to avoid the wild lifestyle of traditional harems. Indeed, use of the quiet strategy increased about 20 percent during a drought year in which the fawn drop had been late and forage production poor. This might be viewed as a survival strategy in hard times because it offers no evolutionary advantage in superior mate selection as do the other two strategies, but it minimizes weight loss, keeping the doe in the best possible condition for a successful pregnancy.

Regardless which mate selection strategy a doe uses, she has to be properly courted the day of her estrus. Bucks less

than three years old don't seem to know this. Either that or they're simply too high on hormones to control themselves. They rush in, skip all foreplay, and try to force themselves on unwilling females. Unless they are absolutely the only male available, they are always rejected. Eventually, perhaps by watching older bucks for several years, they learn proper courtship etiquette. Then, as three- to nine-year-olds, they proceed more cautiously, which saves them and their consorts

After all the ritual behavior, herding and courting, a pronghorn buck will copulate only once with a doe.

energy. A courting buck uses his nose to check each doe for readiness, sniffing her genitalia plus any fresh urine she voids. Like all artiodactyl ruminants, he will lip-curl to push pheromones from the urine into the Jacobson's organ in the roof of his mouth for sampling. When she tests "ready" he announces his desires with a high, thin whine that descends into a low growl. Then he approaches on stiff legs with mane erect, hairs raised over his median gland, and head held high, swinging left and right as if to show off his horns. He sticks out his upper lip and twitches it, simultaneously making a sucking sound that may play on the doe's maternal instincts by mimicking the sound of a suckling fawn. If she tolerates this approach, he'll bump her rump tentatively with his brisket. She may step or even run away, but he presses his suit, repeating the actions and calls as often as

necessary until she stands for him.

Finally, for his long months of guarding and herding, his weeks of patient courting, he is rewarded with a few seconds of copulation and that's it. Each female copulates one time per year. Hardly seems worth the effort, but nature is efficient. This reproduction business is essential, but when you are a prey animal you can't get too emotionally involved. Remain alert, do the job, and get back to surviving. The next generation depends on that.

At this point it appears we have male and female pronghorns playing the mating game at cross purposes. The bucks wear themselves to a frazzle trying to control either vast chunks of countryside or big bunches of females, while those same females often cuckold their would-be masters by switching between harems pretty much at will and picking who will father their fawns. (They usually have twins after their first birth.) What's the sense of this behavior?

To understand the significance of "sampler" and "inciter" female mate selection strategies, one must consider the evolutionary design and needs of pronghorns, then consider how genetic selection influences those needs.

Obviously, nature has designed and perfected the pronghorn as an open-country survivor. Its light pelage blends with the pale colors of plains and deserts.

THE RUT

Its large, wide-set eyes spot danger at great distance. Its remarkable anatomy permits it to outrun every predator except viruses and bacteria. Those, then, are the characteristics that have been bred into millions of generations of this species, attributes that somehow must be selected and perpetuated in both females and males.

Selective breeding is a highly developed animal husbandry science, understood by most high school students. If you wish to develop a race of squat, muscular cattle for maximum beef production, you breed the most squat, muscular bull in your herd to the most squat, muscular cows. You do not breed the long-legged, long-horned bull that beat up all the other bulls. Nor do you allow every bull to breed. Only the best. So it must be with pronghorns. They must breed speed to speed. But, in the absence of a farmer, who selects the fastest, hardiest, most vigorous females and bucks to breed?

Gray wolves, for one. At least they used to before humans erased them from the plains. And millions of years before they went on the attack even larger canids like dire wolves pressed the evolving pronghorn toward ever faster running. Those that couldn't outrun the predators became dinner. Those that could became parents. At the same time, viruses and bacteria weeded out the sick; drought and food shortages trimmed the inefficient browsers, those whose metabolism did not wring maximum energy from every mouthful. These and similar survival pressures honed bucks and does in roughly equal measure, constantly refining the species to be all that it could be. But there is one more test, one final opportunity to fine-tune genetic selection: female choice in mates.

A doe pronghorn has a limited number of opportunities to pass on her genes. She'll bear perhaps 20 fawns over a lifetime. A buck, on the other hand, can contribute his chromosomes to 20 fawns each year over as many as eight rutting seasons. Clearly, here is the best chance to optimize quality. Some mechanism must select the best males for breeding and eliminate the also-rans, and that mechanism is female mate selection.

The manner in which pronghorn females choose mates mirrors the predation-induced selection process. When a "sampler" doe settles with a herd buck who maintains the most undisturbed territory or harem, she has selected a male that has proven his speed and endurance by chasing rivals away. "Inciters" do the same thing, pushing bucks to the limit with last-minute chases.

Even while selecting for speed and endurance, these strategies simultaneously select against large size, just the opposite of what bucks try to do. If pronghorn mate selection were decided by bucks, those that could intimidate or best others in combat would do the bulk of the breeding, pass along those size/strength genes, and soon change the species, making it larger. Large is wonderful for winning wrestling matches but has little to recommend it when a wolf is slobbering at your heels. Nor is large body size advantageous in a dry, open environment where forage is limited and where big, tasty prey animals stand out like beer kegs at a fraternity party. Therefore, mate selection by females who assess the vigor and speed of males helps maintain the pronghorn as a relatively small, extremely fast species.

That explains why male pronghorns are not much larger than females. Among most North American ruminants sexual dimorphism (differences between males and females in shape, size, or color patterns) is significant and selected for sexually. A bull moose can outweigh a cow by several hundred pounds. A whitetail buck often weighs 300 pounds to a doe's 150 pounds. Those species depend more on avoidance (living in dense cover) and strength (leaping over deadfalls) than speed to avoid predation. Massive body size and huge antlers among males indicate a genetic trait for maximizing growth, which manifests itself in female offspring as optimum fetal growth and maximum milk production. This gives fawns and calves a rapid growth spurt so they can quickly outgrow their vulnerable stage and thus escape predation.

Pronghorns, however, will stick to speed. As long as breeding age bucks continue to race and chase one another for breeding rights and females continue to choose the fastest and

most vigorous to father their fawns, this unique, all-American animal will remain the fastest mammal on earth.

A pronghorn doe will give birth to perhaps 20 fawns during her lifetime.

White-tailed Deer

Catch me if you can

Life in the country is pleasant, slow, quiet. Sparrows sing softly from the fence line, geese glide into the pond, and every summer evening a delicate doe and her two dainty fawns slip from the woods to feed in the alfalfa field. Peace. Then the leaves fall and all hell breaks loose.

Two weeks before Thanksgiving the fawns show up alone, nervous, slightly dazed. That evening their mother races across the alfalfa, leaps the fence, tears through the back yard, tongue lolling, and plunges head first through the lilac bushes. And right behind her comes a huge buck no one has ever seen before, his neck as thick as an Angus bull's, antlers like a rack of broom handles, mouth open, grunting like a pig. Before anyone can speak, another buck bursts from the woods, and behind him yet another. Like kids playing follow-the-leader, they cross the field, leap the fence, and dive through the lilacs, breaking limbs and scattering dried leaves.

That night a neighbor rams another buck with his pickup — or, rather, the buck rams him, bumbling into the front fender with his head down as if blind, deaf and dumb. The next Monday a buck wanders into town and jumps through the plate-glass window of Johnson's store. On Tuesday Martha Hinderman calls the fire department to report two bucks in her front yard, stuck together at the antlers!

Welcome to the whitetail rut.

Describing the whitetail as another North American deer is like describing Elvis as another pop singer. The whitetail isn't *a* deer, it's *the* deer, an archetype and an icon. It is grace and innocence, joy and beauty, caution and flight, freedom and wilderness. As a nation, spiritually and physically, we have been nurtured by this animal. It has fed us and sheltered us, lured us and propelled us across most of a continent. Phylogenetically it is the original common deer of North America, a born-in-the-USA, red, white and blue, all-American. Elk might be more regal, moose more massive, caribou more exotic, but, in the hearts and minds of millions, when you say whitetail, you've said deer.

As befits an archetype, the whitetail behaves like most of its close and distant cousins, behavior aimed at one thing — dominion. Yes, crazy as it sounds, whitetails are power hungry. They don't want liberty and justice for all, or peace and love. They don't want to teach the world to sing in perfect harmony. They simply want to be Number One. They want to out-compete, out-breed, and out-produce everything else. Why? Because that's the game, the genetic game of life. He who replicates the most genes wins. Despite their soft brown eyes and fawning nature, whitetails are competitive and selfish.

Don't misunderstand. Our deer are not plotting against us. Theirs is not a conscious choice. Mostly it's instinct. Whitetails are hardware; genes are the software that runs them. They are genetically programmed to be selfish, to grow antlers, to rub trees, urinate on their legs, grunt like pigs, and mate in autumn. Virtually everything they do has been genetically encoded into them

Mature bucks in the upper Midwest or Canada can enter the rut weighing more than 300 pounds. Facing Page — Giant-racked whitetails have three advantages — they can frighten other bucks from competing for does, defeat competitors, and actually be selected by does for breeding.

over millions of years by trial and error. Physical attributes and behavior that work are reproduced to work again. Those that fail are weeded from the population. Whitetails have two options: compete or die.

For females, the competition is relatively easy. Eat and don't get eaten. Come fall, some buck will find you, impregnate you, and, if you can keep eating, you'll produce a couple of healthy fawns. It's a different story for bucks. Because they constitute about 50 percent of the population at birth and just one of them can service many does, most bucks are superfluous. For them, competition rages year-round, and they try everything to win including visual, audio, and olfactory display, bluff, secrecy, fighting and eating.

Eating? Surprisingly, pigging out may be a buck's best offense in the race for space at the annual Mating Ball. As a rule the biggest, strongest, healthiest

males get to plant the most seeds for the next year's crop of fawns. They literally throw their weight around to get what they want, so the more weight they have, the more they get. For example, a mature buck in the upper Midwest or Alberta can enter the rut weighing more than 300 pounds. Good nutrition also produces the best antlers, and large antlers are more important to whitetails than fast cars are to teenage boys. Given plenty of fresh, nutritious browse, a prime, mature whitetail five to eight years of age can, over the summer, decorate his cranium with a pair of 30-inch beams six inches in circumference that spread two feet apart and sport five to eight tines per side. It takes supreme

Sparring between bucks establishes a pecking order. Two bucks of similar size may resort to serious pushing and shoving.

confidence or supreme stupidity to argue with an enemy brandishing that kind of armament. Impressive though they may be, whitetail antlers are but a crown of twigs compared to the headgear of caribou, even considering body-to-antler ratio. Why the difference? Why can whitetails mate and thrive with so much less relative antler mass than caribou? Because they live and hide in the woods.

Whitetails are hiders. They live in dense brush and woods and, although they dash away quickly, are not designed for long-distance running. Instead they accelerate to put distance and cover between themselves and predators, then they loop back and hide. Their fawns hide, too. That gives their dams the luxury of producing relatively small, helpless twins that they can then slowly raise

on fairly low-quality milk. This frees them from the need for superior forage during pregnancy and even after, increasing the likelihood that they'll successfully raise young under adverse conditions. But what does this have to do with buck antlers?

According to the best research to date, antler size in males equates with milk production and quality in females of the same species. The bigger the antlers, the richer the milk. The richer the milk, the more precocious the calves. Caribou bulls grow the largest antlers in relation to body mass of any deer and, sure enough, caribou cows produce the richest milk of any deer. Their calves practically hit the ground running. This implies sexual selection for large antlers. Somehow the males with the largest antlers are being chosen to do the breed-

ing. But how? Three possibilities come to mind: The largest antlered males might rattle their sabers, so to speak, and frighten the sexual urges right out of all other bucks in the neighborhood. Alternatively, bucks with the biggest antlers could thrash all competitors and/or chase them away from most does. Lastly, does could shop around and refuse to couple with any but the largest-antlered buck in the woods.

Researchers have found no "smoking gun" to indicate whitetail does alone determine which bucks get lucky based on the size of their antlers. Rather, it appears sexual selection for antler mass in whitetails results from a combination of the three aforementioned possibilities.

First, bucks begin winnowing the annual winners and losers shortly after they shed antler velvet in early September. Rising testosterone levels make them feisty, and they begin sparring. Two bucks will touch antlers and tussle, feeling each other out to determine who is top dog. This is particularly important for young bucks, who require the tactile exercise to develop an understanding of their relative antler size and strength. That's why immature bucks spar more than mature bucks. Their antlers grow and change dramatically from year to year, from spikes and forkhorns to small eight-pointers to medium eight- or 10-pointers. After the age of three they seem able to assess rank just by looking at other bucks, so they spar less. Most sparring is between bucks one age-class apart.

Sparring lasts for several weeks and sometimes grows heated. Two bucks of similar size might progress to some serious pushing and shoving before they sort out their relationship. Sometimes tines are broken, resulting in immediate demotion. Bucks that lose a sparring match often seek to thrash a smaller buck to reassure themselves of their position in the hierarchy. When the dust finally clears, nearly every buck in the neighborhood knows his place in the pecking order. This spares them a lot of

Serious buck fights, often over an estrus doe, may last for an hour or more. Every year a small percentage of these fights end with the combatants' antlers locked. Unless freed by humans, such bucks are doomed. Commonly, fights end with bucks suffering torn ears or gouged eyes (right).

energy and pain when females reach their heat period (estrus). Coincidentally, they build up their neck muscles for upcoming fights.

Not surprisingly, testosterone levels play a significant role in determining rank. Mature bucks generate considerably more of this aggression-producing hormone than do young ones, thus their tempers usually match their bulk, which is considerable. Aside from mature bodies that are several times larger in frame and muscle than a young deer's, master bucks develop massive necks much the way weight lifters develop bulk when on steroids — except the buck's steroids are

natural. Only fools or suicidal idiots will confront a superior buck overdosed on testosterone.

Sparring for rank isn't foolproof. Matched bucks might wrestle time and again and never agree on who's superior. Should they spar later over an estrus doe, the confrontation could escalate dangerously. In addition, energized males launch extended forays beyond their normal territories to find more females, markedly increasing their odds for bumping into strange, untested bucks. If the two are evenly matched, one throws down the gauntlet and the other, if he thinks he's buck enough,

picks it up.

Now begins a fight over property rights, the property being the doe. Each combatant gives his opponent opportunity to withdraw, encouraging him with a bit of bravado and bluff. Each erects his body hair to enhance overall size. Then he lays his ears back, lowers his head, and walks stiff-legged toward his antagonist. He might stop and paw like a bull challenging a matador. If that doesn't give his enemy pause, he might grunt-snort, hiss, and beat the sap out of a bush. As a final bluff, each walks nearly parallel with the other, angling closer, showing his full body profile. Finally, one charges. Decision time. Flight or fight. If the other runs, he's chased for several yards at antler point, and gored if possible. If he turns to meet the attack, the two crash together and immediately try to throw one another over.

Most of these dominance fights are short-lived, the weaker buck sensing quickly that he is outclassed and had better disengage and cut his losses. *Let*

him have the doe. There are plenty of others in the woods. Occasionally, however, neither warrior will cry uncle, and the battle rages. Legs are spread fore and aft for maximum traction. Grunting, panting, slobbering, and sometimes bleeding, the duo duels. Some battles rage for an hour or more. Each year a few gladiators find themselves caught in a fatal embrace, antlers hopelessly locked, the two condemned to death by thirst, hunger, exhaustion, or predation. Bucks have been found bearing the grisly head of a former adversary, the bulk of the rival having been consumed by coyotes.

Rubs are one way that white-tailed bucks advertise themselves. It is both a visual and olfactory marker. Rubs are usually found on brush and small saplings, but even old fence posts (right) are not immune.

Few bucks escape the rut without enduring cuts, scrapes and puncture wounds about the face and neck. Torn ears and gouged eyes are common injuries. Now and then one buck succeeds in killing another. An exhausted loser of a big fight might find himself suddenly challenged by a lesser buck eager to kick him while he's down and boost his own status. Two or more small bucks have been known to gang up on a fatigued old veteran and ruthlessly gore him to death. Nature does not pity the weak.

In this manner, sparring and fighting are natural-selection factors that determine body and antler mass, but what determines antler shape? Why do whitetail antlers look like whitetail antlers and not mule deer, elk, or moose antlers? Obviously, genes program the architecture, but what started it? Some scientists hypothesize that physical barriers such as dense brush and high winds did the initial sculpting. The whitetail's plow-like antler shape enables it to push through tangled branches, while an Alaska moose's winged palms plane in the tundra wind. Others think offensive and defensive fighting advantages are the more likely design catalysts. The first whitetail that produced beams that curved in, for instance, might have been able to grasp his opponent's rather straight antlers and pull him over, then go on to breed more does and pass on the gene for curved antlers. Similarly, the first buck to sprout supplementary tines might have been able to defeat challengers with simple beams, and that trait would have multiplied.

Regardless how they began, it's easy to see what today's whitetail antlers are designed to do — grasp. The forward-projecting main beams and accessory tines all angle in like claws or hands, the perfect conformation for grabbing and holding an opponent's antlers so that he can be pushed, pulled, and wrestled to the ground. Brow tines set near the antler bases are a last line of defense to prevent an enemy's tines from reaching the vulnerable eyes. Although these antlers can kill, that is clearly not their main purpose. Killing as a way to reduce infraspecific competition for limited resources (females) is genetically self-limiting, too costly for the species as a whole. If males were armed with purely offensive weapons, they would have to abandon fighting (see the chapter on mountain goats) or risk killing each other at an alarming rate, thus defeating the purpose of natural selection.

Fortunately, sparring, displaying, and fighting aren't the only options for determining breeding rights. Equally important, probably more important, is communication — visual, auditory, and olfactory — particularly olfactory. Living in dense habitat where visibility is limited, whitetails have evolved a heady mix of chemical messages aimed at the nose, some efficiently

Scrapes are a buck's way of finding does. At left, a buck licks an overhanging branch, and in the middle photo a buck paws a scrape. The finished product (bottom) may be found each fall in any woods that whitetails inhabit.

combined with visual cues. Two of the most significant are rubs and scrapes, long subjects of speculation and hypothesizing. A rub is a tree, usually a sapling one or two inches in diameter, that has been debarked vertically for one to three feet near its base. Many an amateur naturalist has proudly pointed to one of these, saying "Here's where a buck rubbed the velvet off his antlers." Wrong.

Velvet is stripped within a matter of hours in late summer, often on bushes and shrubs. Weeks later, velvet but a dim memory, bucks are still vigorously pushing and rubbing their antlers up and down saplings. Why? There is no evidence they need to "polish" their antlers, another popular amateur theory — most rubbing is done with only the knobby base of the antlers and the forehead — but there is growing scientific proof that they need to advertise themselves. Rubs, it appears, are combination informational and warning signs along the whitetail road system, the equivalent of: *Master Buck Zone. Proceed at Your Own Risk.* Rubs are not No Trespassing signs, but they do mark a buck's stomping grounds. They are made primarily by mature bucks and are often clustered near popular feeding zones where does and bucks will see them. Their effect is to alert females that a hearty male is on call and to dampen the libido of competitors.

For example, a young male, high on testosterone, probably fresh from routing a forkhorn and overly impressed with his inflated sense of self-worth, is strutting through the woods, dreaming of does. He rounds a corner, and flash! There's a bright white stick staring him in the face. And beside it another. And beyond three more. *Uh oh.* His tail droops and he deflates a bit. Nervously glancing this way and that, he eases up to the first rub, puts his nose to it, and

If a doe finds a scrape and likes what she smells, she'll urinate in it. If the buck returns before her pheromones evaporate, he'll track her down like a bloodhound.

reads the olfactory message its maker has deposited with his forehead glands: *Old buck, big buck, confident buck, selfish buck.*

While saplings without horizontal branches are preferred for rubbing, in some places large trees, utility poles and even fence posts are used year after year as traditional rubbing posts. Farmers and linemen have actually been forced to replace posts and poles nearly worn through by repeated antler attacks.

Like rubs, scrapes are a product of maturity. The older the buck, the more likely he is to rub and scrape. Each act is independent of the other, but both involve scent marking. A scrape is almost always made beneath a slender, overhanging branch or twig about five feet above the ground and usually along a runway or at the edge of a field or meadow. The buck will take this branch in his mouth and chew it like a teething baby does, maybe biting off its end. He may run it over his nostrils, daub it with his preorbital gland, and twist it with his antlers. Then, on the ground beneath it he paws a one- to three-foot oval space down to bare dirt, steps into it and urinates, running the stream over the tarsal glands on the insides of his rear legs.

Exactly what he's created is known only to whitetails, but we have some good ideas. One theory holds that scrapes are like ads in the personals column: *Single whitetail male, 270 lbs., large antlers, seeks adventuresome females for flirting, frolicking, more! My place or yours.* Depending on how randy he is, a buck may leave dozens of these advertisements scattered over several square miles of breeding territory. If a doe stumbles across one and likes what she reads, she may urinate in it before walking off. If the buck returns before her pheromones evaporate, he puts nose to ground and, grunting eagerly, tracks her down so fast it would drive a bloodhound into group therapy. Her trail is left by scent from interdigital glands between her hooves. Such olfactory tracking leads bucks to strange places at strange times. They become so engrossed in following the scent trail that they forget usual caution, often finding themselves in the middle of empty fields at high noon or colliding with a pickup truck at midnight.

This entire scenario has caused researchers to wonder if does are selecting superior mates based on unique chemical signals bucks deposit in the scrapes. Research to date is inconclusive. For one thing, doe behavior isn't consistent at scrapes. For another, scrape behavior isn't consistent among bucks. Some scrape more than others, young ones not at all, though young bucks can and do impregnate does, though usually not when there is a bigger male available. Researchers Larry Marchinton and Gerald Moore at the University of Georgia

WHITE-TAILED DEER RUT: FACTS ON FILE

Family:	Cervidae
Genus:	Odocoileus
Species:	virginianus
Weight:	Highly variable by region and sub-species. Northern males 150-350 lbs., females 100-175 lbs.
Size:	36-42 inches at shoulder.
Antlers:	Single beam curving out, then forward with accessory tines rising vertically from it. Four to 8 tines per side.
Rut:	Late October to early December in north, variable in south into January.
Habitat:	Brushland, young woods, farmland mixed with wood lots, river bottoms and lakeshores, swamps, marshes, forests.
Range:	Central Alberta to South America. Every state in U.S. but Alaska, possibly Nevada, California and Utah.
Society:	Doe/fawn groups in summer, extended doe/fawn clans in winter in north. Bachelor buck groups in summer, solo in fall, mixed herds in winter in north and plains.

suggest that clusters of scrapes mark a buck's primary breeding territory in which he feels most dominant and where other bucks feel intimidated. U of G studies indicate that young bucks are sexually subdued in the presence of mature bucks and as a result do not scrape. By transferring his odors to scrapes and limbs, a boss buck may be extending his dominance and shutting down his competition, in effect being in two places — or dozens — at once. Nevertheless, subdominant bucks regularly trespass on dominant buck scrapes and scent mark the overhanging branches. They have not, however, been known to urinate in the scrape itself. None of this proves they disregard the master buck's warnings. The odors they detect may cause a physiological dampening effect that reduces their urge to breed.

The overhanging branch found at

A whitetail buck checks a doe for estrus. If she is not ready for breeding, he'll leave because waiting will cost him a chance at other does.

nearly all scrapes has a significance way out of proportion to its size. Though seldom larger than a pencil, such twigs are essential for scrape function. Without them scrapes are almost never made. If they are removed, the scrape beneath them will be abandoned. If they are moved to a new location, a new scrape will be made below them. Researchers in Michigan have hung single, slender, drooping twigs over deer trails and consistently induced scraping at the site. Even in the off season bucks will lick, nuzzle, and rub preorbital glands on these twigs without scraping. They are then called licking branches.

It is possible that the limb is a bulletin board for buck-to-buck communication and the scrape strictly an advertisement for does. Another possibility is that the sight and smells of scrapes prompt does to reach estrus early and uniformly. Controlled studies of penned deer have shown that does kept with bucks started their heat cycles eight days

Does will often run if a buck that weighs twice her size charges in. Mature bucks have learned to court the does before breeding.

earlier than does penned alone. Given our weak sense of smell, we may never know the rest of the story. The deer do, and that's what counts.

Some animal behaviorists have questioned why whitetails, separated as they are by foliage, are not more vocal, like elk. A bit of shouting would surely alert potential mates to one's location and availability. Yet, the most you can expect from a whitetail is an alarm snort, some soft grunting, and an aggressive snort-wheeze — nothing to get potential partners together. Why not? Probably fear of predators. Elk during rut are secure in numbers and individual size, so they can afford to bugle and be obvious. Whitetail bucks, however, are small, solitary and vulnerable. They might call for love and, when coyotes show up, discover they dialed a wrong number. Also, because bucks court one doe at a time,

they gain nothing by alerting every competitor in the woods. Fighting over a doe is tough work. Why invite it?

If anyone would benefit from a noisy courtship, it would be the doe. With two or more bucks vying for her, she could easily choose the best. Actually, she already does shop around — in two ways. First, she keeps tabs on local bucks by noting their rubs and scrapes. Then, as she nears estrus she begins spending the bulk of her time in the heart of her territory, a place the master buck of the area is most likely to search for her. Boy meets girl.

Second, if the phone doesn't ring at home (he might be with another doe), she puts on her headiest perfume and steps out to paint the countryside red. This accounts for some of the unusual, reckless behavior often witnessed in November. She roams wide, runs free, dances in a few scrapes, and drives the boys wild. If pressed by an unlikely suitor, she'll lead him on a merry chase, cutting the trail of numerous other bucks, sometimes pulling a caravan of lusting bucks. Fights break out. Girl gets boy. Big boy. Pick of the crop.

These chases might be what is keeping whitetail bucks from evolving into moose-sized animals. If bucks alone decided who breeds, only the biggest, strongest bucks would pass on their genes. Over time that would convert our little deer into a race of giants. What's preventing this? It may be that does, by inspiring chases, weed out the grotesquely heavy males who can't keep up with the lithe little speedsters. In that way bucks would be forced to compromise, to balance fighting size against running ability, and they seem to have done just that. At least so far. This theory raises interesting possibilities about neck size. Bucks may have evolved the ability to suddenly grow huge, muscular necks during the rut for superior fighting ability, then quickly lose that extra bulk so they can better run from predators during the rest of the year. This is pure conjecture. No studies have been done on the subject.

A buck doesn't waste much time on a doe not ready to mate. She will be ripe for no more than 36 hours, but if he hangs around too long waiting for that magic moment, he'll miss other does. He needs a sign of her condition and he

Courting behavior ends with copulation, which may last only seconds. There may be two or three copulations per estrus.

gets it, not surprisingly, from sex pheromones released in her vaginal secretions. When she urinates, some of those secretions are carried to the ground where he can test them. But first he has to get her to urinate. He does this by following her around like an infatuated school boy, grunting like a pig, and holding his head low and stretched forward to keep his nose as close to her rump as possible. At unexpected moments he'll loudly snort-wheeze, apparently in an attempt to startle his consort into urinating. When she finally complies, he quickly walks to the spot and with his tongue transfers a sample to his vomeronasal gland (Jacobson organ) located close to his front lip on his toothless upper palate. Then he tilts his head up 45 degrees, curls back his lip in a Flehmen response, and stands for several seconds as if mesmerized by a haze of intoxicating odors. Actually he is not smelling the chemicals so much as tasting them for analysis. The chemical message tells him if she's nearing estrus.

Meanwhile, if uninterested in her suitor, the doe may do what whitetails do best to avoid unwanted attention — she runs away and hides. However, if she is almost ready to tumble, she'll merely trot a short way or prance in a circle. The buck stays with her and begins serious courting, chasing other bucks away and often chasing the doe to a quiet part of the neighborhood where they won't be disturbed during their short honeymoon. The pair often winds up in atypical places such as suburban back yards, roadside ditches, barren pastures — any place where other deer are least likely to wander.

Delicate, 125-pound does are not in the habit of allowing 250-pound brutes to touch them, so a courting buck has to be cautious and coy until his nose tells him she is actually in heat and ready to breed. Young males don't understand this and charge in like the proverbial bull in a china shop, spooking the doe into a long run. Experienced old hands ease up from behind, grunting sweet nothings, keeping their massive necks and heads low, their chins pointed for-ward in a stereotypical "low-stretch" posture common to all deer, gradually working closer until the doe will stand. Actual copulation is over in seconds. There may be two or three couplings per estrus. Then the relationship is terminated until next year, if the buck still lives. The rigors of rutting — fighting, wounds, all-night carousing, chasing and fasting — exact a heavy toll. Bucks that burn their candles at both ends may lose 25 percent of their pre-rut weight. They seldom live more than 10 or 12 years. But, if they've lived well, their genes will be running the woods and fields in dozens of fawns, grandfawns and great-grandfawns. That is their legacy and reward for taking the selfish time and trouble to be Number One.

Mountain Goat

Love on the rocks

In contemporary parlance, a male mountain goat is a wimp at lovemaking. Here is 300 pounds of muscle and sinew that winters at 10,000 feet, laughs in the face of -40°F blizzards, climbs vertical mountain walls as nonchalantly as humans ride elevators, sleeps on two-foot-wide ledges 2,000 feet above eternity — and crawls on his belly to beg permission to woo.

Uncharacteristic deportment? Perhaps, but eminently sensible, even daring considering the object of the billy's affections. A nanny mountain goat is a shrew — ornery, belligerent, intolerant, and exceptionally aggressive. She's the black widow spider of ruminants. She won't kill and eat her mate after he's completed his procreational duties, but she might poke, stab, gore, or skewer him into an early grave during the preliminaries. With her attitude, it's amazing the species is even able to reproduce. Yet, that intolerant, aggressive nature is the very core of mountain goat survival.

As is the case with all ruminants, sexual behavior among mountain goats is in large part a product of their social behavior, which is in turn a product of their habitat, and mountain goat habitat is stark, hard, cold, and mostly vertical.

Researchers from Alaska to Montana have discovered that goats spend the vast majority of their time on slopes between 40 and 90 degrees — essentially cliffs and the ledges and benches associated with them.

Why has this shaggy beast chosen a life of hardship amid high, relatively unproductive granite peaks? To paraphrase a mountain climber, because they are there. Animals are constantly competing for resources. Some succeed as generalists, adapting to change and prepared to eat nearly anything, tolerate nearly any habitat. The white-tailed deer is a prime example. Others exploit unusual niches and evolve into specialists. Such is the mountain goat, which is not really a goat at all, but a more primitive Rupicaprid, a tribe of hooved mammals from which true goats and sheep evolved. Unlike sheep and goats, which sport thick skulls and heavy horns for butting, the five remaining species in the Rupicaprid tribe — the goral and two serows from Asia, the chamois from Europe, and North America's *Oreamnos americanus* — have thin skulls and relatively straight, short, lightweight horns designed for deadly impaling. Mountain goats do not butt; they stab.

After finding itself trapped in North

THE RUT

America when a post-glacial warming trend melted glaciers and inundated the Bering land bridge between Asia and Alaska, the Rupicaprid ancestor of today's mountain goat faced tough choices. It could evolve predator-avoidance tactics, foraging techniques, and reproductive strategies to compete with open-country herd grazers like sheep and bison, or it could become a specialist, sort of the brain surgeon of the ruminant world. There isn't a huge demand for neurosurgery compared to, say, plumbing, but then there aren't many primates with the special tools and training to handle it, either. Thus, brain surgeons face minimum competition and make enough money to survive and send their kids to college. Similarly,

Mountain goats evolved as animals that exploit terrain that is often uninhabitable for other species. They literally live on the edge.

mountain cliffs in cold northern latitudes don't produce much vegetable forage, but there aren't many ruminants with the ability to reach it, either. This ensures that mountain goats find enough to eat and survive to send their kids to the peaks of higher learning.

So while herd grazers competed on more gentle slopes, North America's new immigrant and only Rupicaprid began taking advantage of neglected forage on precipitous ledges and cliffs. Here it not only reaped an untapped resource but also avoided most predators. As this behavior continued over thousands of years, individuals that climbed better than others and spent most of their time on the ledges survived to produce young that inherited those same characteristics. Inevitably, nature selected traits like strong forelegs and shoulders for pulling and climbing; a phlegmatic, cautious,

plodding nature to prevent missteps; a thick, wooly pelage to survive brutal cold; convex, spongy inner hooves with rough texture for gripping rocks and ice; and an intolerant, aggressive nature to enable each animal to defend limited space and food for its own use. That bad attitude is the key to understanding the rutting behavior of this unique animal.

Living literally on the edge where one false step spells "it's been nice to know you," where the next mouthful of forage might be two ledges up, and where a gentle Christmas snowfall might be the genesis of the avalanche that buries you, nice guys finish last. Survival of the fittest here is no academic plati-

Natural selection among mountain goats has created a species with strong forelegs and shoulders for climbing, a plodding nature, a thick, wooly coat, spongy inner hooves, and an intolerant, aggressive attitude.

tude. There is precious little resilience in the hard rock habitat of the mountain goat, no built-in fudge factor. A white-tail can wander off on the wrong path and merely end up in strange country, but a mountain goat that takes the wrong fork in the ledge could end in free fall. A bison cow can abandon a choice patch of grass to a bigger cow and find another a few yards away, but a mountain goat that gives up a cluster of wild strawberry leaves might go to bed hungry, lose the kid she's carrying, even starve. In short, a goat can't afford to be altruistic.

What advantage would a healthy, strong nanny bestow to the population as a whole if she stepped aside and allowed a weaker nanny to feed on limited tufts of cured grass clinging to a winter ledge? Chances are the fragile female would die anyway, and the con-

siderate, robust one, having given up her meal, might starve with her. If the less hardy one did pull through the cold season, she would pass on genetic traits that made her frail in the first place, thus further reducing the species' fitness. That's a recipe for extinction. It's also why the herding instinct, prevalent among open-country ruminants like sheep, bison, and pronghorns, is largely undeveloped among mountain goats. They don't need the predator protection that herding confers on individuals. The vertical terrain in which they live stands like a moat between them and most wolves, bears, coyotes and mountain lions. What four-footed carnivore wants to wrestle its lunch on a six-inch ledge? Once a goat has reached a cliff face, only the golden eagle can threaten it, and then only if it's young, sick or incredibly

Typical mountain goat society consists of small family groups — usually a nanny with her kid.

stupid. Also, in this vertical land consisting mostly of rock there simply isn't enough concentrated forage to support vast numbers of grazers/browsers. Heck, there's barely enough room to stand. Imagine if two dozen animals followed a browsing herd leader onto a narrow ledge that terminated in an abyss. The guide would eat most of the forage as she progressed. Those trailing would get nothing but frustrated and hungry. Then, when the head nanny hit the dead end, she'd turn around to face a major traffic jam.

So instead of cooperating and sharing, mountain goats squabble and compete with a classic me-first attitude in small family groups — usually just a nanny and her kid of the year, perhaps a few yearlings or another maternal pair. Several family groups will combine from time to time and the youngsters tumble together in summer play, but these asso-

ciations rarely last long. Billies roam singly or in pairs and trios, but they'll join in bachelor bands of as many as eight and stay together with minimum friction most of a summer, leaving the best grazing ranges to the nanny groups. By keeping apart from females with young, they also reduce fighting and potential wounding. Only during winter and the November rut do they mingle consistently with the opposite sex, which are much shorter tempered and more prone to violence.

Any goat ill-mannered enough to approach within about eight feet of another's head, what it considers its personal space, is subject to a pointed rebuff. Researcher Douglas Chadwick once watched an adult nanny in Montana force a younger goat off a 15-foot cliff. He saw males rear onto their hind legs and throw horned punches at one another. He witnessed a nanny chase another to a drop-off and hook her violently four times before the victim chanced a dangerous leap to a thin ledge below. In Glacier National Park one goat knocked another into a raging mountain river.

Through long and consistent observation Chadwick learned that when goats were in

bands with more than eight or 10 animals, they were almost constantly squabbling. One of them could be caught in as many as two dozen altercations in a single hour. During 4,400 hours of observing mountain goats, Chadwick saw 291 aggressive social interactions in dangerously steep terrain. Thirty-nine led to falls. Of those, 18 were physically pushed over the edge, another 18 were forced to make a do-or-die leap, the remainder, innocent bystanders, bumped into trouble with other fighting goats or were the aggressors themselves.

All this violence is due to the intolerant nature of the beast combined

MOUNTAIN GOAT RUT: FACTS ON FILE

Family: Bovidae
Genus: Oreamnos
Species: americanus
Weight: Males 250-300 lbs., females 200-250 lbs.
Height: 36-40-inches at shoulder hump.
Horns: 9-12 inches in old animals, both sexes.
Rut: Late October to early December.
Habitat: Mountain cliffs, benches, rock slides and associated meadows.
Range: Southeast Alaska coast mountains south to Washington Cascades. Rocky Mountains from central Yukon to central Idaho. Introduced to high peaks as far south as Colorado, east to South Dakota's Black Hills.
Society: Nanny/kid family units, small family groups in summer. Solitary billies or small bachelor bands. Mixed associations in winter.

with its strict hierarchical social organization. Essentially, the bigger, older and more aggressive the goat, the higher its standing on the social ladder. Kids of the year assume the status of their mothers. If an adult nanny wants to lie down where a two-year-old is reclining, she marches over to take the spot. If the younger doesn't yield, it risks a horn in the rump. Once displaced, the smaller animal invariably rousts its nearest inferior, and the dominoes begin to tumble throughout the band. Bothersome as these interruptions are, few goats tolerate them without at least token resistance. Because goats do not have the distinctive horn or body size differences of most ruminants, individuals cannot easily tell who is top dog, so they contest the issue. When a goat joins a group, it will test and be tested by every member of that band, even if it had associated with the bunch a day or even a few hours earlier. When two bands meet, the fur really hits the fan.

Considering the lethality of goat horns, all this bickering could have serious repercussions — like major injury or death. To minimize such dire consequences, the species has evolved well-understood body language and ritual fighting behavior. An aggressor may merely stare an opponent into submission. If that doesn't work she might walk or trot toward her target in what scientists term a rush-threat, perhaps pounding front hooves on the ground, growling, and tossing her horns in a threatening swipe that is called a weapon-threat. After a bluff horn toss or two of its own, the victim of such threats usually yields by turning, walking, or trotting away or, when in serious trouble, crouching and

slinking off, perhaps even urinating as the ultimate gesture of submission. Less often, a put-upon she-goat will resist by turning broadside to the threat, standing tall on stiff legs, arching her back, and sucking in her gut to emphasize her shoulder hump and overall body size in what biologists call a classic present-threat. The head is held low, chin down, horns aimed toward the antagonist. This is a serious pose and usually disarms the aggressor.

Into this contentious society enter the males. And none too eagerly, after a summer of solitude and relative peace. But by late October their hormones apparently begin to master them, initiating one of the oddest, most hesitant mating rituals among North American mammals. Mature billies (three-year-olds and up) begin wandering widely to find female bands. At first the old boys merely stand and stare from 100 feet away or more, often for many minutes, even hours. Then they stride away purposefully to gawk and moon over a new bunch of beauties, inching closer day by day and, in their window-shopping delirium, forgetting to eat for most of a month. These voyeuristic voyagers will push through considerable snow and cross the highest ramparts to find isolated nanny groups. This is the species' insurance against inbreeding as well as a test of male fitness. Billies strong enough to hike to the most females potentially pass on the most genes.

But why don't these guys stride right up and check those nannies for readi-

The billies strong enough to hike to the most females will likely breed the most nannies. Note how this billy's coat has been dirtied by wallowing in mud.

ness? Why the long, lonely looks? It could be the billies are screwing up their courage, but more likely they're wisely letting the gals slowly get used to them. The distant presence of virile males might even induce a physiological response in females that reduces their aggression or prompts earlier estrus — the Rupicaprid equivalent of candy and flowers. Regardless, only a fool would rush up to a nasty nanny with nine-inch horns and a prediclection for using them. A protracted, elaborate courtship of a female with a chip on her shoulder is not only wise but mandatory for success. Males that press their suits too eagerly can and will get skewered, and that often takes them out of the ball game — permanently. Researchers have witnessed nannies plunging horn-first into the sides of billies, who almost never retaliate. Most horn puncture wounds suffered by billies are probably inflicted by nannies resisting romance.

Despite the female aversion to physical contact other than for combat, the libido of the males rises with their testosterone levels, leading to pit digging and wallowing. Billies are thought to dig rutting pits as olfactory advertising, as a release of pent-up aggression, or simply as displacement activity since such digging declines as estrus draws near. To excavate a pit, which is actually only a depression a few inches deep covering two to four square feet, a big male sits on his haunches like a dog and begins pawing vigorously with one front hoof after another, throwing dirt onto his belly and flanks (which is why billies are so distinctively dirty in late autumn). A wallowing billy will urinate, perhaps even ejaculate, on himself, then spread these redolent odors in and around the pit. Additionally, he'll swipe the black, crescent-shaped occipital gland at the rear base of his horns across nearby brush and grass stems. Billies may wallow alone or beside another wallowing male in pit-pawing "contests," and several males may use the same pit over time. Sometimes even nannies will hop into a rutting pit and paw, but this could be simply dusting behavior induced by excitement as estrus nears. Biologists have detected no territorial marking significance in rutting pits, but wallowing and its associated odors may stimulate early and uniform ovulation in nannies. This would keep a male in camp until all are impregnated. If estrus were spread over a long period, a billy might lose interest and leave an isolated band to search for other, more receptive females.

Although male goats defend neither territories nor harems, they do protect nannies they are closely attending. Should another male move too near and contest breeding rights, the possessor male literally gets his back up. He will bow up his shoulders, arch his neck sharply downward, and suck in his belly in a dramatic present-threat, but he will turn his head to the side away from the aggressor, just the opposite of what nannies do in their present-threats. It is assumed the shaggy coat, pantaloons, and tall ridge hairs of the goat evolved to accentuate body size during this display. Strangely, the opponent will not attack this huge, vulnerable target. Instead he'll move away or assume a present-threat of his own. Either animal may sidle shoulder-first toward his opponent, try to maneuver uphill, and perhaps roar. The objective is intimidation rather than battle, for neither combatant can defend against his opponent's horns. Unlike deer or sheep, mountain goats have no forked tines or broad horn shields to parry an opponent's thrusts. A frontal attack would severely wound both males, thus reducing their chances for spreading their genes. A billy in present-threat is playing his trump card, and it's usually effective in inspiring the lesser rival to walk away. But when neither male will back down, a mutual present-threat can escalate into the final solution: a ritualistic, rump-stabbing fight.

Two warring billies stand side-by-side, head-to-tail, and begin their dangerous dancing by tramping a tight circle, each tossing horn tips sideways and

For a week or so each November, the Walter Mittys of the ruminant world live their dreams of dominance.

up toward the other's rump, where a strike will get the point across without risking significant injury. This may seem silly to us, but to goats it's both deadly serious and sensible. Their hides are thickest over the rump and rear legs, just as deer's are thickest over the head and neck, where they are most likely to get poked. Most of the blows miss, and usually one gladiator is inspired to widen the circle and trot off. The winner rarely pursues. Sometimes blows are landed, and occasionally one or both males are severely injured. Necropsies of billies wounded during the rut have revealed deep slashes running from shoulder to rump, as many as 10 holes where horns punctured the hide, two-inch cuts across the liver, and severely damaged muscle tissue. No one saw whether those wounds were made by fighting males or ornery females, but male rut fights were suspected.

The Rut

This deadly wounding potential, via natural selection, has made the male mountain goat the unabashed wimp he is. Aggressive billies that assert themselves rarely survive to achieve their end. Appeasing billies, however, do live to mate and pass on their humble, fawning courtship behavior, including a strong tendency to display before other males rather than fight, plus a reluctance to strike with their horns. This is why females, immature males, and even yearlings can easily chase off the biggest and potentially baddest billies early in the rut. It is also a big reason why billies live alone or in bachelor groups most of the year.

Before a male mountain goat is sexually mature, he is just as belligerent and stab-happy as any female, but once he understands the potential joys of sex, he is torn between love and war. Half the time he wants to smack some upstart nanny for infringing on his personal space, the other half he wants to lick her. Unable to resolve the conflicts or establish a dependable dominance rank within the mixed-sex group, he chooses self-imposed exile, joining a bachelor group. This ultimately benefits all mountain goats. There is less stress in both camps and less physical damage to all parties. Should males meet with females before or after the rut, however, they will assert dominance to gain access to mineral licks or other limited resources, reconfirming the wisdom of separate living quarters.

By the second week in November billies with lust on their minds will have crept within 50 feet of the objects of their desire. Immature, inexperienced two-year-olds under the influence of testosterone for the first time will have clumsily rushed and courted the nannies since mid-October, only to be chased, poked, and speared mercilessly. Older nannies especially will have nothing to do with these "teenagers." Experienced males know better and continue their judicious courtship, testing the wind for telltale pheromones, and cautiously closing the gap. When the stinky old geezers finally get around to actively courting, they approach the nannies with the infamous low-stretch, practically crawling on their bellies, which natural selection has cleverly crafted to mimic the submissive retreat posture of a subdominant female. This permits the big males to get close without frightening the nannies away or inciting an attack.

A billy in low-stretch crouches until his belly nearly drags the earth, stretches his neck and head parallel to the ground on plane with his back, tips his horns back, and rapidly approaches the female from her rear, sometimes jerking his head from side to side, flicking his tongue from an open mouth and buzzing softly. The head and tongue motions combined with the billy's prominent beard are hypothesized to visually distract the female from noticing and shying away from the male's large, potentially threatening body. This low-key approach is diametrically opposite the male's present-threat in which body size is maximized in broadside display and the head/beard hidden. If the nanny isn't ready, she will ignore the male, trot away, or reject him forcibly with a rush-threat. If she is ready, she will permit the male to sniff or lick beneath her tail. He might then lip curl or kick a front leg swiftly between or along the nanny's haunches. If she stands for this, he may attempt a mount, clasping his front legs around her haunches and burying his nose in the thick fur on her back. Although nannies have been seen to permit three mountings in a single 48-hour estrus period, it is not known if they actually copulate more than once.

When females enter their heat and permit mounting, billies finally feel safe to assert themselves. No longer meek and obsequious, they rush from one nanny to another, sniffing boldly, licking, lip curling, and kicking. They paw bedded nannies to make them stand, resolutely present-threat if challenged. Young, inexperienced females seem frightened by the suddenly confident males. If rushed, they crouch submissively, flatten their ears, urinate to distract their pursuer, and flee while the big boy lip curls. Master billies no

longer take any guff from juveniles or two-year-old males. Instead, they drive them to the edge of the band. Most of these youngsters give up and quit rutting entirely.

It's a short hiatus, but for a week to 10 days in late November, during the bulk of the breeding, the Walter Mittys of the ruminant world finally live their heady dreams of dominance. By mid-December, however, it's back to the same old same old. Nannies are again violently resisting billy courtship, nasty as ever. Some males slip away to be alone, others take their chances in wintering areas used by the nannies and kids, though not in close association. All are probably relieved to have survived another bout of procreation, wondering what ever possessed them to indulge, and blissfully unaware that, come next November, a hormonal process they don't understand will compel them to do it all over again.

Elk

When I'm calling you

If Hollywood ever starts handing out awards for rutting performances, elk will be perennial favorites to win Best Picture. The elk rut has it all — larger-than-life actors, dramatic scenery and sound, suspense, action, and a twisting, turning plot. Elk versus elk, elk versus nature, elk versus himself in an ongoing saga of intrigue, deceit, suspense, and survival, opening this September in a wilderness near you.

Although the North American elk, a close relative of the European red deer and a recent immigrant from Siberia, is neither as widespread nor common as the whitetail, its rutting behavior is probably more widely known. Instead of sneaking about woods and fields in isolated pairs saying little more than a grunt, elk parade their autumnal rituals in sizable herds through open meadows and grasslands, squealing, chirping and bugling their enthusiasm. From the redwoods of California to the hardwoods of Wisconsin, from the great Rocky Mountain national parks of Canada to the dry pine forests of Arizona's Mogollon Rim, elk are passionate and proud of it.

Indeed, a bull elk has something to be proud of. The branched, six-point antlers adorning a mature wapiti (a Native American word for elk), are one of the most symmetrical, regal crowns in nature. Any bull macho enough to grow and carry such a rack has a right to trumpet about it. More than that, he has a genetic duty to show off those regal bones to impress cows , intimidate other bulls, and use them as weapons against both bulls and cows in an effort to pass on his royal inheritance. Antlers are one of the keys to natural selection among elk.

Elk rut routine is pretty straight forward on the surface. Stag parties underway since November begin to break up in mid-July amid rising testosterone and tempers. Suddenly old Bill and Harry don't seem so appealing anymore. Increasingly belligerent individuals wander off to strip velvet and polish new weapons. They spar with old friends, then attack saplings, thrashing them into submission, rubbing bark into the rough pearling encircling their antler bases, matting forehead hairs with tangy sap while trying to remember what's been missing from their lives since last October.

Cows! That's it! They forgot all about those lovely cows.. No wonder the guys were starting to wear thin. *Cows! Where are they?* Giddy as high schoolers on

THE RUT

prom night, bulls in late August begin cruising the woods and screaming their anticipation. If they find a spring or damp spot of ground they flop into it, tearing at soil and surrounding vegetation with their antlers, rolling and caking themselves in mud. Then they stand and bugle again, finishing the long, high whistled notes with deep grunting as they simultaneously anoint their belly and neck hairs with a redolent stream of eau du urine.

Searching, bugling as often as four times a minute, advertising himself in odor and song, the bull hunts and parades until he finds a herd of

Rising testosterone cause bull elk bachelor groups to break up amidst sparring. Old friends become rivals.
Facing page — A giant bull licks and grooms a cow.
Spread, next page — A bull bugles to remind his harem he's the one they want.

unclaimed cows or they find him. He approaches, head laid back, bugling, grunting, spraying himself with a fresh layer of perfume. *Here I am you lucky ladies. All yours.* He circles the band, staking his claim. From here on out it's merely a matter of keeping the gals from straying until each is ready to breed. Piece of cake.

But soon a satellite bull shows up, makes a run at a cow lingering at the edge of the herd, bugles. *Why that little upstart.* The herd master roars, squeals, grunts, and trots toward the would-be thief who stares wide-eyed at the master's rack, turns tail and runs. But he doesn't vacate the area. Instead, he hangs at the edges, bugling when he feels safe, sneaking as close as he dares, hoping for another chance at a cow. More wandering teenage bulls, unable to attract and hold harems of their own, join him, occasionally sparring with one another

to work off tension, always hoping to steal a moment of romance.

Days pass. Weeks. The bull and his collection of cows are comfortable with one another. He bugles regularly to let them know he's on guard, herds them by sticking his nose in the air, laying his antlers over his back and rushing them. He mercilessly chases any that attempt to stray, brutishly poking them back into the fold with a tine or two, but he pretty much follows where they lead, too busy guarding and displaying to eat much. The females know his scent, his voice. To date none have reached estrus, but he courts them daily just in case, moving up from behind them with his antlers held high, his tongue flicking in and out to show he means no harm. If one allows him to reach her, he licks and

A herd bull meets a challenger in a furious elk battle. Brute strength, balance and endurance are tested.

grooms her as if he were a doting mother attending a calf. When a cow urinates he sniffs the discharge and lip curls to sample pheromones with the Jacobson organ in the roof of his mouth. All is on schedule and going well. Only a few more days and breeding begins. Oh boy! What was he thinking all summer hanging out with the guys?

Then real competition shows up. A huge, swaggering, 800-pound stranger. Roaring and squealing, he comes, leaning his head expertly side-to-side, slipping his massive antlers between pine trunks without so much as nicking one. What a rack! Seven tines per side. The main beams stretch back five feet and spread so wide you could drop-kick a grizzly between them broadside.

The herd bull swaggers out to meet the threat, his head tilted back, bugling for all he's worth, spraying himself. The challenger stops and bugles back. The

herd bull steps to a sapling and beats it soundly. His rival does the same. Both urine spray again, extremely confident. Neither will back down. They approach one another, turn parallel and march in lock step for a hundred yards, showing their respective bulk and antlers to one another. At the end of the march both turn suddenly, sweeping their huge racks low over the ground before marching back the other way. Still, neither yields. Suddenly the enemy senses an opening and whirls. The harem master spins antlers forward and catches the attack in a crashing of bone. Tines mesh with tines, antlers hold, and the shoving match begins.

His antlers are his gold chains and sports car, his urine-soaked hair his cologne, his bugle his loud mouth. He's the guy at the bar you'd like to punch out. For elk, it works.

Now it is brute strength, balance, and endurance. They push, shove, slobber, and grunt, tear the moist meadow soil as hooves strain for purchase. Five minutes. Fifteen. Finally, after 20 minutes of exhausting battle, the antagonist disengages, spins instantly and runs, but not quickly enough. The herd bull sticks a tine into his enemy's rump and draws blood, then chases him from the field before stopping to bugle and urine spray.

His Lordship has retained possession of his cows, but at a price. His neck is gouged and punctured in five places, one eye is gone, and now there is a teenage bull running wild through the harem. The old warrior will rest a minute, then rout the youngster and restore order.

Such is the obvious drama of the elk rut, but what lies behind the scenes? What is the plot and why have the actors been given these lines, these actions? The story is a basic survival epic. The players must overcome nature — cold, deprivation, disease, predators, the evil that lurks within their own genes — to perpetuate their species. Each bull, each cow has the responsibility to carry on the legacy of millions of generations before them. They must mate and produce not just their replacements, but the finest replacements possible.

The reason bulls command center stage in this show, full of themselves, seemingly the star attraction, is because they are. Cow elk progress through an inconspicuous estrus, giving almost no indication of their condition. The bull, on the other hand, is a sex symbol, Hollywood's leading man, a walking, talking, stinking billboard hawking his own virility. His antlers are his gold chains and sports car, his urine-soaked hair his cologne, his bugle his bragging, loud mouth. He's the guy at the bar you'd like to punch out — except you know you can't. He's also the guy who gets the girls.

Such a character might not sound like God's gift to womankind, but for elk it works. A bull's job is to attract attention to himself and earn breeding rights by guarding and protecting cows from other bulls. That's why he's a pompous, stinking braggart. The more cows he protects and the better security he provides for them, the more calves he will sire. The cows' job is to select the most vigorous mate. At least that's what biologists assume. They haven't proven conclusively that cows do the selecting, but circumstantial evidence is strong.

To win the mating game, a bull first has to grow antlers, the bigger the bet-

Cows and calves summer at lower elevations to avoid predators. Actually, the browse is better at higher elevations where the bulls spend the summer. There, they put on weight and grow antlers.

ter. Simultaneously he must store enough fat to fuel himself through the rut with a bit left over to keep him alive while he recuperates and starts eating again. Those two nutritional needs explain former mysteries about summering bull elk. For instance, bulls seek shade and wallow in mud holes more often than do cows because they are producing one calorie of heat for each calorie of fat they are storing. Bulls live alone or in small bachelor bands so they don't have to compete against cows and calves for forage. Biologists once believed males behaved altruistically, leaving the best feeding sites to the cows and calves, but they've since learned that the eating is actually better where the bulls hang out. The reason cows summer in herds in more open habitats at lower elevations is to avoid predators. Free from the onerous chore of producing antlers and extra rutting fat, they can

afford to sacrifice forage quality for security. Not having youngsters to protect, bulls can reverse that, which explains why they generally live at higher elevations in mid-summer. Vegetation begins growing later there and is thus more tender and nutritious. And why do bulls wander more than females? To find isolated pockets of young broadleaf plants rich in minerals and protein. Such plants sprout in the aftermath of forest fires, landslides, uprooted trees and similar unpredictable disturbances.

Interestingly, after the rut bulls winter alone and often high in the mountains for similar reasons — there is little competition for food, they can more effectively hide from predators, and they can outrun shorter-legged predators in deep snow. In open country, however, bulls collect in large "stags only" herds to gain traditional predator safety in numbers. There are more eyes, ears, and noses to detect stalkers, but more importantly there are more bodies to deflect attention. A lone bull has to outrun all the wolves. In a herd he has only to outrun one other elk. In a herd of cows, however, a large, antlered bull would stand out like a neon Eat Here sign. The federal government had better not force gender integration in these elk's clubs.

Depending on his age and health, a bull nurtures his growing antlers for three to five months, relatively defenseless in the meanwhile. That's quite a risk. Is it worth it? Yes. A bull's head gear is his signature of health, vigor, and competence. It is a sexual characteristic more than a weapon and proves he can avoid predators, find nutritious and abundant food, and efficiently metabolize raw forage. A balanced, symmetrical rack mirrors overall health. Lopsided antlers are grown by injured or sick bulls. In the game of love, a bull without antlers is like a peacock without a tail.

ELK RUT: FACTS ON FILE

Family: Cervidae
Genus: Cervus
Species: elephus
Weight: Males 600-1000 lbs., females 500-650 lbs.
Height: 4-5 feet at shoulder.
Antlers: Main beams 4-5 feet long, up to 5-foot inside spread, 5-8 tines.
Rut: Late August through mid-October.
Habitat: Mix of brush, grassland, meadows, forest.
Range: Historically most of North America, now mostly western mountains from north British Columbia to Arizona. Re-introduced populations in several eastern states.
Society: Cow/calf groups in summer, large herds in winter. Bachelor bull groups in summer, small groups to large herds in winter.

THE RUT

Once his rack is fully formed, between late July and mid-August, a bull must strip it of velvet — which he does in a matter of hours, sometimes days. Then he tests it against social equals in sparring matches, an elk version of human arm wrestling. Two bulls nearly matched in size approach one another with antlers lowered, slowly waving side to side. Often they yelp to signal their intent to spar. Then they touch horns almost gently and begin pushing and shoving, chirping and squealing as they do. This provides valuable training for youngsters and enables all bulls to correlate their size and strength against others so they'll recognize more formidable opponents when they see them and avoid deadly fights in the heat of passion. The ability to discriminate between medium and high-ranking antlers is a social skill learned over three or four years. Young red deer stags, raised experimentally without a prime male to teach them their proper place, ignorantly and rather stupidly attacked a much larger, superior stag when he was introduced to their pen. That doesn't happen in the wild.

To interrupt sparring, one bull disengages and avoids eye contact with the other, who stands and waits, proof that this is a game. To end the match, one contestant jumps back, prancing and head weaving like a playful puppy. During spirited sparring, and more often during all-out fighting, tines and whole antlers are sometimes broken, a serious loss. Regardless of his age, body size, or previous ranking, a bull that loses his head gear loses his status. Demoralized, he'll retire from breeding for that season. No bugling, no urine spraying. Research has shown that these debased bulls often grow smaller racks the next season. Talk about an inferiority complex.

The stable social hierarchy established through sparring lasts until antlers drop in late winter. At that time, no longer sure who's on first, bachelor bands break up until enough new antler growth can be produced to re-determine rank. Teenagers and yearlings don't shed their racks until April or May. A healthy, mature bull may shed his old crown in March, so you can bet he moves out. What would be worse than a bunch of teenagers pushing you around? Immediately after antler drop the pedicel scars over and a new antler begins to sprout. During velvet stage testosterone levels are low, estrogen high. Males are essentially feminized, so there isn't much fighting. Disputes are settled cow-style, with flailing hooves so that tender antlers aren't damaged. One researcher discovered that high levels of vitamin A present in green vegetation also reduces aggression.

Unwilling to trust antlers alone to deliver his macho message of self worth, a prime bull adds a dash — some would say overdose — of olfactory stimuli, most of it urine. He has the uncommon ability to aim this discharge in a stream between his forelegs to hit his neck and belly. At a wallow he will douse himself, then roll in the mud and rub his neck on the sides of the pit. Subsequently he might rise to rub and horn a tree, transferring scent from his neck and mane onto it.

Urine dousing is also associated with thrashing, during which brush, grass or saplings are slashed and torn with the antlers in a dominance display that is simultaneously noisy, odoriferous, and visual. As such, it is often performed by a herd bull when a competitor approaches too near a harem. Some biologists contend this antler thrashing is vestigial scent-marking behavior harking back to the distant past when antlers were skin-covered peduncles (similar to the "horns" of giraffes) designed to transmit sex pheromones. In support of that hypothesis, they point out that antler velvet today contains the highest concentration of sebaceous glands (oil-producing — the same type that give us pimples) in deer skin. That could explain why elk spend so much time rubbing their foreheads and antler bases on saplings, then thrashing those (scent-

Elk douse themselves in a wallow, roll in the mud and rub their neck in the pit. The bull may then rub and horn a tree, transferring scent from his neck and mane onto it.

ed?) stems with their antlers. Or why they stir their urine-soaked wallows with their antlers. Whitetails carefully rub scent from their preorbital glands (slits in the skin below the eyes) and forehead glands on branches before antler thrashing them. Mule deer do much the same thing. Clearly, something is going on, and our infamous inability to detect subtle odors might be preventing us from sniffing it out.

A bonus to thrashing is the occasional collection of debris in the antlers, the equivalent of a man wearing elevator shoes. Bulls seem to realize their size has been augmented, and they parade proudly to show off and intimidate rivals even further. Researchers have watched red deer use brush-adorned antlers to suddenly rush other stags in an unritualized fashion, as if knowing they have the visual upper hand. Indeed, no stag so attacked has been seen to hold

his ground.

Bugling is the *coup de theatre* of a bull's multi-media advertising campaign. From a distance it sounds like a fine, lovely whistle rising up the scale. It isn't. Up close the complete call is harsh and threatening. It begins with a low-pitched roar that quickly rises several octaves to a murderous scream that plunges to a series of throaty grunts or chuckles tied to belly twitching and urine spraying. The low roar resonates from the deep chest of a mature bull and is designed for local listeners only. It does not carry far. The high scream, however, carries

The elk bugle begins with a low roar that quickly rises to a murderous scream. The roar is designed for local listeners only.

remarkably well across open plains. The European red deer advertises itself with a low-pitched roar, quite adequate for close-range work in the woodlands in which it evolved. Elk, however, had to change their tune as they evolved across the open expanses of Siberia and Alaska. Only in the past century have elk been pushed by man into vestigial forests. If kept there long enough, their ringing bugles might eventually evolve back to a low pitch.

Given the male-dominance bias of our culture, it's only natural that we assume bugling is an intimidation or challenge to other bulls. After all, when a satellite bull bugles, a harem master

will bugle back. But he'll also bugle when no other male is around, which seems a waste of energy. Such indiscriminate calling would only attract unwanted attention. If, on the other hand, we assume bugling is designed to attract females, why should a herd bull bugle and risk tolling in competition? He already has his groupies. What's going on?

While a herd master's wails probably do warn smaller bulls to keep their distance, their real value is in attracting and holding cows. The reason a bull continues to bugle after he's collected a harem is to out-perform other males. If he were to shut up, his harem might question his masculinity and switch allegiance. The bigger and more confident the bull, the more eagerly he will call, knowing that he can defend what is his. Any outsider brave enough to sound off near the harem will be hunted down and roundly thrashed if necessary. Likewise, a small bull temporarily in charge of a harem will keep his mouth shut and attempt to move his bevy away from a bigger, tougher-sounding bull, but to no avail. The cows will see through his ruse and soon dump him. After all, they heard the big Lothario singing in the wings. How you gonna keep 'em down on the farm ...?

An experienced old bull also uses his bugle to reinforce a cow's faith in him, to reassure her that he isn't as violent as he looks and sounds. While a young bull will rush a cow repeatedly and try to mount her against her will, the herd bull will approach her slowly while telegraphing his intentions. He moves up from her rear, holding his antlers straight up (where she can see them rather than low over his back as he does when aggressively herding), and flicks his tongue to show that he is going to lick her. If she doesn't appreciate this foreplay, she will run away shaking her head and champing

The rutting behavoir of elk has evolved into a remarkably effective way of choosing Mr. Right.

her jaws as if she were silently chewing out the old oaf. That's her way of saying "I've got a headache." And the bull, considerate, gentle giant that he is, defers by turning away from her and bugling. To a human observer this sounds like a scream of frustration (and who knows, it might be), but the cow, like one of Pavlov's dogs, soon associates his bugle with his considerate behavior. Ergo, to avoid unwanted sexual advances, stick with the devil you know.

Another valid inquiry is why a bull continues to spray himself with urine even after he's amassed a gaggle of adoring females. First, it shows his extreme self-confidence and signals downwind competitors that he's still on duty.

Second, it may stimulate a physiological response in his cows, initiating an early and uniform heat. Among herding ruminants in northern climates, where spring green-up starts with a bang, the ideal is to birth collectively in as short a span as possible to catch the vegetative peak and to "swamp" predators. The sudden appearance of hundreds of calves might attract undue attention, but while local predators digest one or two, the others are growing larger and stronger. Soon they will be able to keep up with the herd and outrun tooth and fang. If the young were born over several weeks, they would become nature's version of Burger King — fast, convenient food on demand.

All things considered, elk rutting behavior is remarkably effective for selecting Mr. Right. It is not, however, perfect. Although sparring and displaying prevent much unnecessary strife, bulls suffer significant casualties. Dr. Valerius Geist, zoologist and professor at the University of Calgary, Alberta, figures the average mature bull will take 50 cuts and punctures about the head and body each rut. Even spike bulls, largely bench warmers in this game, suffer about 30 wounds each per rut. In fact, a spike was the battle scar winner in Geist's research — it absorbed 70 hits heavy enough to register on its hide. Elk suffer such wounds willingly as the price of love, but sometimes they pay the ultimate fee — life. Geist estimates a 10 percent rut-related mortality among most antlered species. Despite their defensive design, elk antlers are also offensive weapons, and when a fighting bull can gain an advantage over a rival, he jumps at it, goring and spearing in a bloody frenzy. A bull that stumbles, trips, or falls and consequently cannot parry his opponent's thrusts is in big trouble. Recently an amateur photographer at a national park witnessed a large bull antler-thrash a children's swing in a park. The swing's chains tangled in the bull's antlers, effectively rendering him defenseless. A teenage bull, recognizing

The elk rut is not without casualties. Researchers estimate that mature bulls suffer 50 cuts and punctures each year and a 10 percent rut-related mortality.

a rare opportunity, attacked and gored the prime bull to death.

For all its posturing and ritual, all its grace and beauty, the annual elk rut is still survival of the fittest, a wild, passionate, primitive, no-holds-barred contest. Bulls that choose to play it run themselves ragged displaying, calling, and trying to herd a band of fickle, wandering cows. Almost none are able to hold a harem throughout the entire rut. The best they can hope for is to control the most cows for the greatest number of days during the peak of the breeding, roughly a two-week period. This gives them their best shot at fathering calves. Overly zealous bulls that begin the rut too soon, push themselves too hard, and take too many chances die young. A bull that paces himself, waits until cows are in estrus, and minimizes fighting can survive to breed over several seasons and

thus pass on more of his genes. But few bulls seem capable of such plotting. Their hormones are too powerful, their lust too great. For most it's all or nothing. By the time the festivities are ended in mid-October they will have lost nearly all their body fat, including much stored within their bone marrow. If deep snow comes early, these wounded, exhausted, depleted animals die of malnutrition. Cow elk generally live into their late teens. A bull rarely makes it to 13. If you've a hankering to bet on an elk surviving winter, don't bet on a herd bull.

The payoff for this tremendous effort, this sacrifice, is a better breed of wapiti. Essentially the rut is nature's extended quality-assurance program. Before a bull is qualified to breed he must live long enough to produce large antlers. That proves he's disease hardy and predator smart. He knows how to stay out of trouble and how to find plenty to eat when times are tough. If he can grow exceptional antlers and lay up enough fat to fuel himself through a month or more of rutting, he has the genes future bulls and cows will need to maximize growth and milk production and to lay up summer fat against hard winters when forage is scarce. If he can out-call, out-stink, out-wrestle, and out-compete other bulls, he has strength and endurance. His legacy will live on.

Sheep

Boys will be boys

Few creatures in nature are more magnificent than a mature, wild male sheep perched on the ramparts of a snow-capped mountain. With massive horns curling about his head like a regal crown, a ram is the symbol of strength and nobility. Snow white Dall sheep, black and white Stone sheep, chocolate brown Rocky Mountain bighorn sheep — regardless the subspecies, sheep are the essence of wilderness, masters of all they survey.

Behaviorally, wild sheep do not live up to their public image. They are more selfish than noble. They epitomize the "me first" attitude. If they were humans they'd all be in jail, juvenile detention centers, or psychiatric wards, and it's the fault of those huge horns. Because of those horns, wild sheep virtually never grow up.

North American sheep have done with horns what caribou have done with antlers — taken them to their extreme, changed a simple tool into an elaborate symbol of rank and power. The more youthful and vigorous the ram, the faster and larger his horns grow. The bigger a ram's horns, the higher his rank. The higher his rank, the more ewes he breeds. And the more ewes he breeds, the more lambs he sires that carry his genes for

huge, fast-growing horns and the juvenile behavior that goes with them. It's a self-fulfilling prophecy. As long as habitat and forage remain productive, sheep remain youthful, vigorous, aggressive, large, and behaviorally underdeveloped.

Contrary to old wisdom, sheep horns did not evolve as weapons. They are too inefficient for deadly combat. A mountain goat can do more damage with 10-inch stiletto horns than a Stone ram can do with 50 inches of curled horn the diameter of a small tree trunk. Ram horns are better formed for defense, but they didn't evolve for that purpose, either. Ewes, with their small, thin horns, can absorb blows from other ewes and even mature rams without injury by letting the force push their heads and necks downward while the energy dissipates through tough neck muscles and tendons. Bison minimize head butts with a dense mat of forehead hair. Even wild hogs are able to ram head-to-head hornless, thanks to thick, pneumatic skulls.

Because sheep horns dissipate some body heat, a few biologists have speculated that they may have evolved as thermoregulatory devices. That idea is bankrupt by the simple fact that ewes live with rams and don't overheat despite their much smaller horns. Also, large-horned sheep species around the world live primarily in cold, northern climates. Who needs to unload BTUs at 20 degrees below zero?

What does make sense is that a ram with a natural tendency to butt will dominate his peers if he is blessed with slightly heavier butting horns. This lifts him to the top of the pecking order where he has increased opportunity to pass his big-horn gene to many lambs. Horns, then, didn't evolve as weapons to injure, but as rank symbols to dominate. And that has had unusual repercussions.

Rapid growth is a product of youth and superior nutrition. In most animal species, youngsters grow quickly, reach adult physical stature, then stabilize both physically and behaviorally. In new country (such as that exposed by receding glac-

Both rams and ewes have horns. They did not evolve as weapons, nor as thermoregulatory devices, but as symbols of rank.

iers) where vegetative growth is vigorous and competition low, sheep thrive and extend their juvenile growing period over the bulk of their years. They remain in something of a twilight zone of perpetual adolescence. Under such conditions, rams with youthful tendencies grow bigger horns than rams that mature sooner. Eventually, such vigorous, big-horned rams out-breed mature, smaller-horned rams, and the entire sheep population inherits those juvenile tendencies, including aggression, vigor, and unrefined social skills. The upshot is that today's North American sheep act more like unsupervised teenage boys overdosed on testosterone than noble guardians of the mountain wilderness. And ewes are locked in a permanent juvenile state.

These insights into sheep society were uncovered in the 1960s by Canada's innovative student of animal behavior, Dr. Valerius Geist. During much of that

Thanks to inherited juvenile tendencies, sheep tend to act like unsupervised teenage boys.

decade Geist lived in various parts of the Canadian Rockies, watching and recording Dall, Stone and bighorn sheep behavior. He later reported and masterfully interpreted his observations in several scientific papers and a seminal 1971 book, Mountain Sheep.

As Dr. Geist noted in that book, humans may be repulsed by much of sheep social behavior, but sheep are not people. Our moral conceits and prejudices are not necessarily applicable to the *Ovis* genus. The simple reality is that sheep, not despite of but because of their social system, have thrived for millions of years across the northern hemisphere from western Europe to the Baja Peninsula of North America. They must be doing something right.

The basic rule of sheep society is that all males treat all subordinates as estrus females, which means they behave "homosexually" to subordinate males. This is because females have small horns, which make them look like yearling males. They are essentially "frozen" in their physical development. For 363 days each year they act like juveniles, but during their two-day heat they act just like young rams. Or do young rams act like estrus ewes? This becomes one of those chicken-before-the-egg puzzles. To avoid confusion, dominant rams treat all sheep as if they were ewes in heat which, surprisingly, results in a stable social environment. At least until the ewes really do come into heat. Then it's mass confusion as everyone tries to mate with everyone else.

When a ram is between two and three years old it begins feeling its hormonal oats and pestering the ewes with which it has been living since birth. They have neither time nor energy to waste on such shenanigans, so they begin avoiding pesky young males who soon tire of the unresponsive females and begin to wander. Eventually they run across an older ram, which they immediately follow. They are sheep, after all. Thus begins the youngster's introduction to masculine sheep society. For two or three years he will associate with various older rams, learning from them where to move seasonally for summer forage, pre-rut staging activities, the rut itself, post-rut feeding areas, early spring feeding grounds, and possibly a salt lick or two. Depending on how many old rams a young recruit follows during these formative years, he may inherit up to seven distinctly different seasonal ranges to which he remains quite faithful, even though this condemns him to some ridiculously wasteful commuting habits. A summer ram, for instance,

In sheep society, all males treat subordinates as estrus females.

might migrate 10 miles south to a pre-rut range when there is a perfectly good one two miles north. Then he might reverse himself and move 15 miles north to a rutting ground, passing through two breedable flocks of ewes on the way. This isn't all bad, however. It reduces inbreeding and disperses hybrid vigor throughout the herds.

When a young ram leaves the ewes and joins the brotherhood, he finds himself in a strange society. In any bachelor band of two to 20, the largest horned ram will be the leader. He asserts his dominance by approaching his subjects with a low-stretch threat display — neck and head extended and in line with the back. Then he kicks them with a front leg to the belly, brisket, or haunches. This is exactly how he courts a ewe. For the ultimate gesture of superiority, he mounts the lower-ranking ram.

In response to this ignoble treatment,

Bachelor bands usually number from two to 20. The largest-horned ram is typically the leader.

lesser rams demure, permit the mounting, and sometimes even depress their backs (lordosis) afterward and urinate — the very thing ewes do after being bred. The master ram may test this urine by lip curling — again, the same thing he would do to test a female for estrus. It is the lesser ram's obligation to tolerate this degrading handling, then rub his horns against the face and horns of his transgressor, sort of a "lick my boots" appeasement gesture. This horning behavior is so prevalent that ram facial hair is uniquely pliable and tough to resist it. Dall rams often have dirty faces from such rubbing, which may serve to transfer the dominant's preorbital gland scent onto his associates, creating an olfactory gang identity and preventing unwarranted aggressive action on dark nights.

It almost appears that a subordinate ram's behavior is a desperate attempt to "hide behind a woman's skirts," to avoid being pounded by bigger rams, but dominant rams seldom instigate horn clashes. That is the prerogative of subdominants. As long as inferior males let larger rams have their way with them, no serious fights break out. This is what permits rams to live together in relative peace all winter and summer. If they all behaved like grown-up males, they'd never stop fighting.

Head butting is common, just one ram casually bumping another in play or to make a point. Even clashing — two rams charging into one another head first — can be done playfully. It's the equivalent of sparring among deer. A serious dominance clash can break out anytime one ram thinks it can overpower another and climb the social ladder. If he fails, no hard feelings. He simply returns to his former station. These contests of strength and endurance may last only a round or two, but sometimes they drag on for hours during the rut when two evenly matched rams argue over a willing ewe. This is sheepdom's version of a world class, heavyweight boxing match and every bit as exciting.

When clashing, one ram will maneuver uphill if possible, rise on his hind legs like a rearing horse, and run forward bipedally several yards before dropping to smash the keeled edge of one horn into his adversary. He puts a lot of "body English" into the charge, generating energy with his forward run, his falling body, a sudden push downward with his massive horns, and the concentration of the blow onto one small edge of one horn. Only after that edge has landed will he twist his head to bring the other horn into play, delivering a one-two punch. His abrupt stop often propels his hind legs off the ground. The dominant ram, always on the alert for these challenges, quickly turns and lowers his head to catch the blow,

Dominance clashes occur any time a ram thinks it can overpower another.

kicking up his hind legs and allowing the energy he absorbs to push him downhill, where he adroitly lands. Both combatants step back, lift their heads, turn their horns at a slight angle to their opponent, and freeze for as long as a minute in a present-threat. It looks like they're waiting for the pain to subside, but actually they are showing each other their horns. This allows them to visually correlate the substantial wallop they just endured with the horn that delivered it. In this manner they learn to accurately assess the strength and dominance rank of any ram merely by studying his horns.

If neither combatant feels he was bested by the initial contact, both walk apart, turn, rear up, and take another whack. After that it's clash and present-threat, clash and present-threat until one finally capitulates. If things get really heated, they'll skip the present-threat and get right back to another clash. During serious clashes in summer, blood erupts from under the base of the rapidly growing horns. Frequently blows are not landed perfectly and horns slip past one another,

Rams rise on their hind legs before dropping to all four for the charge.

the rams plunging headfirst into the ground. Noses are broken, hides sliced by passing horn tips, and horn tips themselves cracked and splintered. This fight breakage — and not the intentional rubbing of the horn tip on rocks to clear a ram's line of sight — is what causes brooming or wearing of ram horn tips. This is why one horn is often broomed while the other is not. The broomed horn is the one the ram habitually angles in to deal the first blow. Its tip is most likely to strike the rival's horn, suffer the brunt of the impact, and splinter. Dall sheep, more often than bighorns, have unbroomed horns because they fight less than bighorns. Additionally the ends of their horns flare to the sides, where they are less likely to be struck.

Sometimes rams get so excited or irritated with one another, particularly around an estrus ewe, that they ignore ritual fighting styles and fall to brawling, knocking one another in the ribs, swinging horns left and right, bumping, and shoving with shoulders in a free-for-all. During these wild struggles, horn tips sometimes hook over one another, temporarily locking the antagonists together, which they rarely appreciate. To free themselves they pull, spin, and tussle wildly, out of control. Such a fracas usually ends with the victor escorting the loser from the grounds at full speed, horns prodding all the way.

With large and delicate heads, humans wonder how rams can knock craniums so freely without killing or at least addling one another. The energy of two 250-pound rams connecting after a typical charge is more than 2,000 foot-pounds. How do they survive? Predictably, they have evolved physically to withstand the crash force. Their brains are shielded under two layers of skull bone separated by about two inches of air space honeycombed with strut bones. These struts rise to fuse with the horn core which is itself one-quarter-inch thick. Even the skin covering a ram's forehead and nose is five or six times thicker than that covering his body. A ram doesn't have a head so much as a crash helmet.

Shields and helmets, however, cannot cushion all 2,000 foot-pounds of hammering. The overflow is handled by a broad connection of heavy muscles and tendons between the back of the skull and the vertebrae of the neck. When horns meet, each ram's nose is forced down, allowing the connective tissues supporting the spine to flex much like the leaf springs of a car. The principle is the same as a skydiver bending at the knees to absorb landing impact.

After a summer of carefree feeding, butting, dominating, and being dominated, rams in late September or early October move to their pre-rut grounds for two to five weeks of increased interaction. Here, strangers and old rivals meet, size each other up, and challenge those they think they can beat. They especially like to huddle on ridge tops, all putting their heads to the center to horn and rub the new master. They are spirited and playful. Rules loosen and subordinates take liberties, clashing more often, making threat-jumps toward superiors, even

Rams can become so irritated with each other that they result to brawling. Facing page — The head and horns of a ram are designed like a crash helmet.

mounting those they know they aren't supposed to mount. Five- to seven-year-olds do the most mounting, having reached maximum body size but not yet maximum horn size. They can afford to push their luck with dominants since they have the bulk to give almost as good as they take in a clash. Sometimes things get out of hand and the smallest sheep in the bunch becomes the brunt of a mounting party. One after another the older rams humiliate him until he flees, leading the band on a wild chase across the slopes — just as ewes will do before they are ready to mate.

By the time the pre-rut stag party is over in late October, dominance ranks have been stabilized, but then everyone leaves for different rutting grounds. Here, discipline falls to shambles. Even though two-year-old rams are sexually mature, they remain behavioral teenagers until eight or nine, which means they pester females with juvenile courtship for six or seven years. These animals rut with all the reserve and finesse of a college freshman at his first frat beer party. Gangs of juvenile males molest anestrus (not in heat) ewes, chase them, abduct them, and try to breed them. If one attempts a mount, his buddy slams into his side, horn first, butting him off. He then quickly tries to mount, but another lusty associate bumps him off. The ewe squirts away and leads a

two-mile chase, a half-dozen irrepressible, overzealous, oversexed rams in hot pursuit. She may try to lose them in the cliffs, hide in a crevice, back her rear against a wall, or lie down and cling desperately to the earth. But her persecutors don't care. They kick, prod, pry, and butt her loose, and the chase is on again.

When these boorish good-old-boys clubs can't find a female, they gang up on the youngest male at the party, insulting him by kicking him like a ewe, courting him, sniffing his urine, lip curling, and mounting him.

Under these deplorable conditions, a ewe's only hope is to throw in with a mature male, the biggest-horned, oldest ram she can find, and stick with him. If she is in estrus or very near it, he will guard her, confronting the rivals with

Competition for females is intense. Young rams may chase ewes and fight among themselves while trying to breed.

horn threats and low-stretch approaches with a twist, meaning he twists his horns back and forth around the axis of his neck, drawing attention to them and accentuating their size and potential. He may add an emphatic growl. If he's considerably larger-horned than they, they get the hint and leave. More often they hang out nearby, huddling, knocking heads, and seemingly plotting among themselves while harassing their smallest member. Until things settle down, the ewe must not panic and run, for if she does the young rams will be on her like hounds on a rabbit, and the older, heavier ram may not be able to keep up. Then she must either find her way back or accept the largest of the gang, who will then turn on his buddies and assume the role of dominant, courting male.

Indications are that ewes reject small rams as long as possible to mate with a dominant male, but sometimes, outnum-

bered, they simply give up. Such a sacrificial ewe may be mounted in turn and repeatedly by several randy rams. Geist reported one estrus ewe being mounted 39 times in 88 minutes by seven different rams! Another time a dominant, 12-year-old ram mounted a ewe 11 times in one hour. The next day the old-timer looked a bit tired and a feisty four-year-old was able to spirit the ewe away and mount her 39 times in three hours before the big ram relocated them and took back his property.

Given that most deer copulate a handful of times, some only once per estrus, a natural question is why the seemingly excessive mating among sheep? These characters make rabbits look like underachievers. Once again, juvenile behavior gets most of the blame, that and intense competition, which is fierce for several reasons. First, because horns are not deadly weapons, young rams can afford to take chances. If they are caught, they suffer little worse than a hard punch in the side. A deer caught in the same uncompromising position would get punctured. Second, because sheep do not defend harems or rutting territories, all rams have equal access to ewes and it's every ram for him-

self. Third, virtually every ram on the mountain is young, vigorous, and largely undisciplined. If they want something, they take it. Remember, the rule in this society is to treat every subordinate as an estrus ewe. Fourth, because rams commonly hang out in gangs, there are always plenty to compete and interfere with their friends. Five- to seven-year-old rams have achieved nearly full body size, and their horns are often big enough to give the truly mature rams, eight-

SHEEP RUT: FACTS ON FILE

Family: Bovidae
Genus: Ovis
Species: canadensis (Bighorns), dalli (Dall and Stone)
Weight: Bighorn males 200 to 300 lbs., females 125 to 180 lbs. Dall and Stone males 150 to 250 lbs., females 120 to145 lbs.
Height: Bighorns 36 to 42 inches at shoulder. Dalls and Stones 34 to 40 inches.
Horns: Bighorns: bases 14 to 17 inches circumference, length to 49 inches. Dalls and Stones: bases 13 to 16 inches, length to 50 inches.
Rut: Late October to mid-December.
Habitat: Grasslands on mountain slopes and beaks, broken river breaks, badlands.
Range: Western mountains from Alaska south to Baja, California, east to South Dakota badlands.
Society: Ewes and lambs in small flocks. Rams in bachelor bands from two to twenty. Mixed flocks during rut.

to 12-year-olds, a serious challenge. Couple that with their immature behavior, and no one gets any peace or privacy.

Ultimately, this interference has genetically predisposed sheep to quick copulation, taking "slam, bam, thank ewe ma'am" to new lows. If a ram wants to see

A mature ram will court a female with some reserve. He will lip curl (above) to test for estrus, approach in a low-stretch, lick her, bump her with his chest, and if she's ready, hit her with a front leg kick. If she stands for all this, she's ready to mate.

his genes in the next generation, he'd better plant them fast before one of his buddies knocks his chance away. For this reason rams have evolved physiologically to deliver semen in many small doses rather than one large one, as do bison and most other bovids.

Such quantity-over-quality romance also explains why sheep do not indulge in such common ruminant rutting strategies as calling, scraping, or scent marking. There's no time! Because they live in herds on traditional ranges and in rela-

tively open habitat, they have no trouble finding one another, so forget calling. Ditto for scraping a la white-tailed deer. If a ram wants a ewe, he just looks over the hillside flock and picks one. Since they aren't territorial, there is no need for scent marking, the way pronghorns do, and because they don't try to hold harems, they don't need to perfume themselves, as bull elk do. Sheep may be the least odoriferous rutting ruminant in North America. They've made horns their entire game plan. He who has the biggest horns wins. Most of the time.

Small-horned rams get their chances if they can find an estrus female alone or abduct one and mount her before her former beau catches up. It's a long shot, but worth a try when your worst punishment will likely be a butt to the butt.

Not surprisingly, overt aggression and excitement are sheep aphrodisiacs, their version of foreplay. Mounting is most frequent after a chase, fight, or other ram-to-ram competition. Mature rams alone with willing ewes often seem listless and disinterested. Their courtship is subdued, cautious, sometimes lacking altogether, especially after the first week of breeding. It's as it they're thinking "Gee, without the guys here to beat out, what's the point?" Well, the ewe knows the point, and if she isn't getting it, she takes it upon herself to provide the requisite stimulation. She'll jump suddenly in front of the ram, startling him, run a few paces, whirl, and lower her horns in a threat pose. Then she may approach him in low-stretch, butt him in the side or shoulder, horn his face, rub her entire body across his chest, and back her derriere against him. Get the picture, big boy? If he doesn't respond, she might leap away as if to flee, staging a remarkable pantomime of typical gang rutting behavior.

When a fully mature ram, usually over eight years old, does court a female, he actually does so with some reserve. None of this uninvited mounting. He'll follow her, sniffing, testing her urine, lip curling, approaching in low-stretch-and-twist until she permits him to touch her with his nose. He may lick her anogenital region and bump her with his chest a few times. When he thinks she's ready, he hits her with a front kick to the body. If she stands for this, she's usually ready to mate. Most copulations take place in late November and early December so that lambs are born all at once to swamp predators and take advantage of rich spring grasses.

Once breeding in concluded in early or mid-December, ewes ignore rams and get on with the business of finding enough to eat for themselves and their growing fetuses. Males gradually drift away, rejoin bachelor groups and move to their respective wintering ranges. This sexual segregation is critical because of the rams' relentless testing and poking. Ewes can't afford to waste energy on such distractions. Their objective is to maximize eating and minimize all other activity. This is why females live in small flocks with known associates. They can't even afford the stresses of interacting with strange females, let alone males. Once lambs are born, older rams are a real threat. They treat even these small sheep as potential mates, butting and kicking them just as they would another ram or an estrus ewe.

Both sexes, then, seem satisfied with their segregated lives, the ewes feeding and nurturing their lambs, the rams courting, mounting, and dominating each other — until the decreasing light of late fall signals hormonal changes. Then they're back together for another month or two of their uniquely wild, rampant sex-ploitations that perpetuate one of the most dramatic, regal, noble *looking* hoofed mammals in the world.

Caribou

Love on the run

One hesitates to describe a natural phenomenon as an orgy — the term has such negative, anthropomorphic overtones — but that's what the barren ground caribou rut looks like. Thousands of animals intermingling, grunting, chasing, fighting, and copulating right out in the open. Some might call it shameless, undignified. But they are not the ones who have to mate in the snow at two below zero.

The October rut, brief respite though it is, might be the one joy in a caribou's life. Here is a creature forced to live on the run, homeless, almost constantly migrating from summer tundra to winter taiga forests to find enough to eat. Wherever it goes it is stalked by wolves, plagued by flies, and tormented by maggots that burrow beneath its hide and through its sinuses. For half the year these long-suffering deer endure temperatures as low as -65°F, losing weight the entire time. Then, just when things warm up and they start putting on a little fat, the July bug season begins. For the next six weeks the attack is so relentless that these northernmost members of the Cervidae family again — in the heart of the growing season — lose weight. They seek windy hilltops where flies are blown away. They huddle together, nose to nose with flanks out in an attempt to block their winged predators. They hide for hours on barren gravel bars or snowfields. They snort and kick and suddenly stampede in a desperate effort to escape their tormentors. More than 1,000 warble fly larvae have been counted under the hide of a single caribou.

The amazing thing about these long-suffering animals, aside from the fact they even survive, is that during this debilitating fly season the bulls grow the largest antlers of any living deer (in ratio to body size) —beams eight inches thick, five feet long, broad, palmated tines and as many as 30 points to a side. Why? If southern whitetails can rut and mate with relatively small antlers in a land of perpetual food and mild temperatures, why must caribou labor under 30 pounds of essentially useless bone? Wouldn't they be better off funneling the nutrients wasted on antlers into body fat to see them through winter?

One can't fully appreciate caribou antlers without first meeting a critter that no longer exists, *Megalocerus giganteus*, the largest antlered deer that ever lived, the so-called Irish elk. This moose-sized Cervid, once common across northern Europe, stood seven feet at the shoulder, probably weighed 1,500

pounds, and died out about 11,000 years ago near the end of the last Ice Age. Many say its antlers were its demise, so excessively large that their bearers couldn't afford the costs of feeding them, the energy drain of holding them up, or the burden of carrying them away from predators. With a weight of 100 pounds and a spread of 14 feet, they became the Spanish Armada of antlers, too big and clumsy to work. The biggest Alaskan moose rack is but half that size. Like moose antlers, *Megalocerus* antlers were broadly palmated and rimmed with tines, but the palms and tines curved backward instead of forward, where they would have served a defensive or offensive function. That is significant. It suggests *Megalocerus* antlers were more symbolic than functional and did not lead to its demise.

Now why would an animal go to the trouble of growing oversized antlers that weren't much good for fighting? The answer was proposed back in 1871 by another extinct European, Englishman Charles Darwin, who suggested antlers were the mammalian equivalent of colorful and elaborate bird feathers, symbols of virility, cosmetic baubles to impress females. The buck or bull who could grow the most ostentatious antlers would woo and win the greatest number of females. Could that be? Could something as superficial as cosmetics have driven evolution in Irish elk? Could it be driving survival of the fittest in caribou? Absolutely. More or less.

For such a system to work, the ornamentation must be more than haphazard puffery. It must be visual proof of sub-

Bull caribou grow the largest antlers of any living deer, in ratio to body size.

stantial genetic advantage; it must mirror physiological or behavioral superiority. And that's what antlers do. They are nature's resumé proclaiming their bearer's accomplishments. Because they require so much energy to produce, they announce "I am a survivor, the best of my breed. I have evaded wolves and grizzlies, avoided starvation, shrugged off maggots, laughed at cold, pawed through deep snows, and found forage where others could not. I have lived long and prospered. Will you have my baby?"

What female could resist such an opening line? But do caribou cows actually notice bull antlers, and can they discriminate between the best and the rest? And if they can, do they choose to mate with those superior bulls?

To answer some of those questions, Dr. Anthony Bubenik of Canada led an innovative research study that pitted rut-crazed caribou bulls against humans in caribou suits. The men hung caribou "rump" patches on their backs, strapped mounted caribou heads on their chests, stepped onto the tundra, and essentially shouted "el Toro!" Bulls responded appropriately. If the decoy had a larger rack, they behaved submissively, going so far as to touch noses with the bogus bull, a stereotypical submission gesture. If the researcher positioned the fake head in a threat posture, big bulls attacked. (One assumes the researcher had memorized the "stop-attack" posture before trying this.) Most significantly, females were always attracted to dummies with the largest antlers. In fact, they were so attracted that they would jilt a real bull for the two-legged impostor if it had bigger antlers, despite its human odor. If a cow is willing to pair with a dummy, antler size must be a critical criterion to breeding success in this species. By extension, the same must have been true for *Megalocerus.* The healthiest, most vigorous males grew the biggest antlers, attracted and bred the most cows, and passed the big-antler gene to succeeding generations. If big antlers hadn't worked, they would have been eliminated via natural selection.

The next question is why choose antlers to highlight superior males? Why produce a heap of bones each June only to discard them each November? Couldn't these animals have come up with a less wasteful process to impress the females? Grow them and throw them? Sounds like something Madison Avenue would concoct. Couldn't the biggest male simply kick his subordinates to the back of the pecking order, horse style, and prove that he's the best?

In a word, no. Strength contests are important, but they don't genetically determine certain physiological traits that open-country ruminants like caribou need, namely rapid growth as calves to evade predation. In maturity and health, a caribou can stay one step ahead of wolves. It's a different story when they are babes. There's the bottleneck. If a calf is to survive, it must virtually hit the ground running. The peripatetic herd cannot wait around for each little one to grow up. The longer they stay in one place the more they overgraze the delicate tundra and the more predators they attract. There are few places for a helpless infant to hide on the tundra. Imagine the carnage if 100 cows camped on a barren plain for a month, calling their calves out of hiding several times a day to nurse like whitetail fawns. Wolves

could sit watching and waiting until a calf stood up and rang the dinner bell. Easy pickings. Caribou cows, then, must be genetically selected for their ability to divert nutrients from their own body maintenance into production of a strong, precocious calf and rich milk to nurture its rapid growth. Such youngsters survive to carry on these genetic traits while weaker, slower-growing calves nourish the next generation of wolves. But caribou cows represent only half the gene pool. What about the bull?

According to a popular theory advanced by Dr. Valerius Geist, a bull's antlers are its surrogate baby. Like a real fetus, they demand efficient foraging and funneling of nutrients toward their growth. The larger they grow, the better the bull's genetics for those traits. Thus, when a cow ogles a nice set of antlers and mates with their owner, she is selecting a genetic partner that mirrors her own reproductive ability. The resultant fetus gets a double dose of superior growth genes. *Voila!* Super calf. No twins for these deer. The objective is to produce one big, advanced specimen. Within an hour of birth the calf (up to 20 pounds) is not only standing but grazing. A few hours later it is walking, running, even swimming with the herd. *Adios, wolves, we're out of here.* In four to eight weeks it is already weaned. The big-antler-sex-selection-system works because it correlates antler production in males with fetus and milk production in females, resulting in precocious calves that are up and running in an environ-

What may seem wasteful, the annual growing and shedding of huge antlers by caribou, is actually a process that helps caribou cows select bulls that are genetically superior.

ment where running means survival.

Like today's caribou, yesterday's *Megalocerus* must have been a herding species of open grasslands that stayed one jump ahead of predators by advertising male superiority with gigantic antlers. Its outsized cranial adornments were not symptomatic of genetic degeneration. Quite the contrary, they reflected its superior ability to eat well, run fast, and thrive, just as the caribou does today.

That does not mean that a caribou bull's antlers are all show and no go. An old theory holds that these impressive bones evolved as fighting instruments and, based on the way caribou use them, that's at least partly true. The beams have about 65 times more strength than needed simply to hold them up for show.

Being fast on their feet is a survival technique for caribou. Genetically-superior calves can join the herd within hours of birth.

But they have only slightly more strength than needed to prevent their breaking under the pressures generated by two 500-pound behemoths trying to push each other over. Clearly, they have been carefully designed as instruments of battle. An argument against antlers as primarily fighting tools is that they are excessively ornate for the job. Any relatively short, compact, thick rack would be sufficient to grasp an opponent. Why waste nutrition growing the superfluous tines extant on caribou racks? Why not pour that fuel into more muscles to gain a strength advantage?

One answer is that antlers are more than just tools; they are also symbols of fighting ability. As symbols, their intimidation factor validates ornamentation. If a bull can convince another to back down without fighting, he establishes dominance without risking bodily injury. Only males of similar antler size

actually have to fight to settle the question. Indeed, this is what caribou and other deer do. Yearlings and teenagers spar more than mature males to get a "feel" for various antlers and how they relate to body size and strength. Serious battles almost never occur between large-antlered males and small-antlered males. The big guys merely display their wares and the runts turn tail. So let those tines grow. The more the better. It's an efficient system.

Zoologists are divided over which of the three theories — intimidation, fighting, or attracting females — drives evolution in caribou antlers, but the animals themselves don't waste time debating. They wield

Caribou antlers are more than fighting tools. They help intimidate an opponent, which allows a bull to win a standoff without risking personal injury or expending energy. By early September, bulls unwrap their antlers by thrashing shrubs.

their racks freely for each purpose in what appears to be a wild mating free-for-all that rages for most of a month across the tundra.

It begins with the usual preliminaries. By early September the oldest bulls have piled 40 to 50 pounds of fat over their rumps, and they'll burn every ounce of it within two months. They begin by thrashing their velvet antlers against arctic shrubs or spruces to unwrap their latest creations. Shortening day length has prompted increased testosterone production, and that has made them feisty, though it hasn't yet

driven them from their fraternal clubs. Soon they begin sparring, testing the new hardware, measuring themselves and their competition, establishing a pecking order that will eventually become chaos. By late September the summer-gray necks of mature bulls have turned bright white and sprouted long hairs that hang and sway much like the dewlap of a moose. This is postulated to accentuate what researchers refer to as the head pole gestalt, the overall "look" of the all-important antlered end of the beast. White fur draws the eye, long neck hairs enlarge apparent body size, and tall antlers top it off like a red umbrella in a holiday cocktail.

By late September bands of bulls and cows are beginning to turn south toward their winter ranges in the taiga forests where stunted spruces hung with arboreal lichens provide both wind protection and forage. Some will migrate more than 200 miles, covering 12 to 40 miles per day, breeding along the way. Small bands mingle and become big bands. Big bands merge into small herds. Small herds coalesce into a seething mass of caribou grunting, feeding, sparring, chasing, displaying, and fighting. Researchers have reported seeing a dozen fights under way at one time. In the cacophony and confusion, how does a bull go about securing a willing mate?

Obviously the scrape-marking tactic of the whitetail is a waste of time. Here today, gone tomorrow. An elk-style harem? Impossible. Hundreds of cows coming and going, meeting and mixing — a bull would go crazy trying to maintain order. The cautious courtship of the mule deer? You'd be left in the dust. Calling like moose? What for? You're surrounded by eligible partners. The only viable option seems to be the one taken — aggressive sampling of cows followed by selection of the ripest one.

In early October, as the herd moves, bulls roam freely through a cornucopia of possibilities, using every possible means to win partners. Naturally they display their bulk and antlers, thrashing

Bulls constantly check cows for estrus as the herd migrates in October.

shrubs, and twisting and turning their slightly lowered heads to show off every tine and paddle both to intimidate rivals and excite cows. But they also urinate on their tarsal glands, then spread that scent by rubbing their noses over their hocks. Since mature, confident bulls do most of this scent marking, they become more obvious to the cows, day or night, and probably intimidate smaller bulls simultaneously.

When the herd is moving, bulls constantly scent-check females, following in their wake, nostrils working, lip curling if they get a chance to sample urine. When they detect a possible candidate, they trot alongside her, head held high and the jaw parallel with the ground. Then they turn just their head toward the cow, maybe grunt to get her attention, and give her a good look at their magnificence in a caribou version of a Parisian fashion show. Frontal view, turn

for a broadside look, turn again to show the back. There's that white neck mane swaying, those glorious antlers reaching up and back and curving forward, broad shovel palms over the nose, heavy bez points atop them.

Should another bull intrude at this juncture, he and the displaying male may pause for a brief scuffle, but they can't waste much time or their fight will be futile because the cows are moving on. They dispense with niceties, hit hard, and fight furiously, twisting and shoving. Within 30 seconds at most it's over. The loser quickly moves to check a fresh prospect. No hard feelings. Sometimes both bulls return to the same cow, agreeing to settle their grudge later if it proves necessary. Often a fight will draw a crowd, some of whom will immediately pursue the cow that instigated the fracas. Others will jump into the brawl, many times freeing one of the original warriors to return to courting.

When a bull strikes a hot prospect, he switches to tending behavior. Wither she goest, he goest, displaying and grunting his desire. He ignores nearby fights, but rigorously chases other males away. His goal now is to guard this one cow until he can breed her.

If she stops to rest or feed, he may hunch his hind legs beneath his belly and re-perfume his hocks, rubbing the tarsals together to mix and disperse the cologne. He might also indulge in a long stint of "bush-gazing," an odd behavior no one has quite figured out yet. The bull stands dejectedly, head hanging low, nose almost touching the ground as if staring at a bush. This might last several minutes or several hours. Could these bush-gazers be sleeping on their feet so as to be ready to react instantly should a challenger arrive or their girlfriend run off?

Caribou living south of the tundra use slightly different mating

CARIBOU RUT: FACTS ON FILE

Family:	Cervidae
Genus:	Rangifer
Species:	tarandus
Weight:	Males 300-500 lbs., females 200-300 lbs.
Height:	46-58 inches, depending on sub-species.
Antlers:	Males: Beams 8 inches thick, 5 feet long, spread 5 feet. Females: 80 percent smaller than males'.
Rut:	Early September to early November.
Habitat:	Tundra, taiga forests, boreal forests.
Range:	From Canada border north to Arctic islands, circumpolar. Known as reindeer in Europe and Asia.
Society:	Cow/calf herds in summer, bands in late summer and winter. Bachelor bands in summer. Mixed herds in rut.

strategies than the big barren-ground herds. Because they don't migrate far, if at all, and don't live in great herds, bulls often attempt to maintain harems, though their success is questionable. During a detailed study of a southern population on the Gaspe peninsula in Quebec, A.T. Bergerud of the University of Victoria, British Columbia, documented a rank hierarchy among 26 bulls that rutted atop a flat-topped mountain, the bulk of them staying there for most of a month. Dominant stags tried to keep cows from leaving groups, threatening them and trying to herd them, but they couldn't quite pull it off. As a result, females came and went, pretty much selecting their mates. Nevertheless, a handful of dominant stags bred the bulk of the cows. Bergerud documented a tending behav-

Barren-ground caribou bull with cows during the rut in September.

ior he called "slurping" wherein a bull would approach a cow in low-stretch and flick its tongue in and out much as a bull elk does during its courtship approach. He also witnessed one fight to the death in which a master bull caught his retreating opponent in the side, bowling him over and pinning him to the ground for seven minutes, then rolling him over four times when he tried to rise. The dead bull suffered internal injuries and compound fractures to both left legs, bloody evidence of the brutally serious business of mating among wild caribou.

Depending on the latitude at which a herd is living, actual breeding may begin in early or late October, but 90 percent is completed within 10 days. That is important to the anti-predator strategy of "swamping." If the vast majority of cows can birth their calves at the same time, there will be too many for local wolves to eat. By the time they digest

their first few kills, the remainder will be older, stronger, and better able to escape. Cow/calf tundra herds in July coalesce into herds numbering tens of thousands to simply overwhelm predators by sheer numbers. Nevertheless, the average survival rate of calves in many herds is reported no better than 50 percent in their first 12 months of life. Where wolves and grizzlies are common, 90 percent losses have been documented. Adult mortality, however, is a low 4 to 6 percent.

One last question remains: Why do cow caribou have antlers? The most widely accepted theory to date is that they use them to compete with teenage males for winter food. "Teenage" bulls grow racks about equal in size to a cow's small antlers and, because they aren't involved much in the rut, they linger with cow herds until March. If cows weren't equally armed, they couldn't defend themselves from the antler attacks of these small males, who attempt to chase females from the feeding pits they've dug through snow with their hooves.

By the time the October rut/migration ends, the most active bulls will be nearly exhausted. Procreational duties completed for another year, they shed their antlers in the snow like weary old gunslingers hanging up their holsters, unaware that, 10 months hence, they'll have a sudden urge to come out of retire-

The most widely accepted theory why caribou cows have antlers is their use as protection against "teenage" bulls, which live near cow herds almost until spring.

ment. For the next several weeks, though, they keep pretty much to themselves and concentrate on resting and eating. It's going to be a long, cold winter, and the fun's over.

Bison

Love and the sumo wrestler

Based on their stolid nature, one might imagine watching bison procreate to be as fascinating as watching turtles sleep. One would imagine wrong.

If pronghorns represent sexual ballet among ruminants, bison are sumo wrestling. When a 2,000-pound bull courts a 1,000-pound cow, you don't expect finesse and subtlety. This is power sex. Gravity and mass alone ensure that, particularly when two bulls disagree over who should be waltzing Matilda.

As one would guess from the size and shape of their horns, bison bulls are not into ostentatious display. Forget pageantry and brandishing of elaborate weapons. Either fight or go eat grass. A challenging bull simply walks toward his chosen target and stares. And the lesser bull slips away. But an older bull, say seven or eight years old and full-sized, isn't so easily cowed. If Matilda is on his arm when the herd bully approaches and stares, he'll just stare back. This response might not sit well with its intended receiver, and he'll say so by shaking his head like a dog shedding water. Then he will open his mouth, stick his tongue out, and bellow, a great leonine roaring that carries more than

three miles. His adversary roars back. Now both bulls paw great clouds of dust into the air. One urinates, then flops onto the ground, and wallows as if limbering up, slamming his great hump to the earth and raising even more dust. The other follows suit. More roaring, grunting, pawing, head shaking. The bluffs aren't working. They walk closer and begin nodding their heads up and down quickly, in unison. Suddenly, from 10 yards out they charge.

When two tons of black, hump-backed fury meet head on, dust flies as if beaten from old barroom rugs. Dense mats of forehead hair six inches thick and a heavy, flat frontal skull blunt the initial crash. Broad horns block enemy horns from passing. Locked head-to-head, the great beasts begin shoving, their sloping hindquarters pushing, their great humped backs straining, each trying to throw his adversary's head aside so he might pass and gore the vulnerable underbody. Contrary to what each beast's bulk implies, this is not a slow, grinding process. Earl Drummond, former manager of the Wichita Mountains National Wildlife Refuge in Oklahoma, described rut-fighting bulls "as fast as wildcats and a lot more wicked. They put their heads to the ground and make quick jabs up with the horns, striking each other in the head, neck and shoulders. The hair and wool is *(sic)* flying."

Bison fights are the stuff of legend, epic battles waged on the grand stage of the wide, wild American plains when Indians still ruled the West and Manifest Destiny was just beginning to pillage its way across the continent. In the 1830s, artist George Catlin was on the northern plains where he saw several thousand bison "in a mass, eddying and wheeling about under a cloud of dust ... plunging and butting at each other in the most furious manner."

The rut was well under way, and Catlin was fascinated. "The males are continually following the females, and the whole mass are in constant motion; and all bellowing in deep and hollow sounds; which, mingled all together, appear, at the distance of a mile or two, like the sound of distant thunder."

That thunder can still be heard today in Hayden Valley in Yellowstone National Park. Across this grass and sage mountain valley, free-roaming bison still gather each summer to rut, fight, and mate. Though there are no longer the thousands of Catlin's day, from a hill the plain below still seems to be crawling with dark brown insects as bulls circle and chase among the grazing cows. The crisp mountain air rumbles with grunting and bellowing.

Explorer John Fremont in the 1840s witnessed 20 bulls "butting and goring one another without distinction" in a wild fight that eventually centered on an old, thin bull that was knocked to the ground several times, only to climb up again to defend himself against the herd,

"blinded with rage." The bull would have been killed, in Fremont's estimation, had he and his men not broken up the brawl.

Other explorers, naturalists, and modern researchers have witnessed defeated bulls being chased for hours by a stream of bulls. In confined herds, such beaten warriors are often cornered and finished off.

Why all this brutality and fighting from such a plodding wild cow as the North American bison? Because that's the only way they have of determining who wins breeding rights in an open-country herding society where everyone knows everyone else's business. Whitetails can hide to reduce competition for estrus does. Elk can flash big antlers, bugle, and urine spray to maintain their harems. Sheep establish and maintain a distinct dominance hierarchy throughout the year based on horn size and predictable behavior patterns between subdominant and dominant rams. But bison rut right out in the open in great herds, and they wear no distinct badges of rank. After they reach full adult size at about seven years of age, one bull looks much like another. His social standing then depends on his day-to-day health, vigor, and fighting ability. Bull bison establish dominance only through trial and error, and that's why they interact so much.

The mass battles witnessed by 19th century adventurers were not relic behavior. They still occur where bison roam relatively unhindered in large numbers, such as in Yellowstone National Park. Researchers call them "fighting storms," and they play an important role in the rapidly changing

Stories of bison fights were a huge part of the legends of settling the Old West.
Next page — Dust flies as a pair of one-ton bulls fight for rutting dominance.

social ranking of bulls.

During a fight storm, a strong bull might begin at the top of the heap, winning several matches in quick succession. But as the battle rages he tires, and toward the end he might be bested by what normally would be a weaker male. Researchers have seen a single bull engage in 10 fights during a 22-minute fight storm, winning at first, but tiring and losing later to bulls he had previously beaten.

This conflict pattern is exacerbated by the ever-changing configuration of large bison herds. Males do not remain in stable and consistent groups. They haven't opportunities to establish cohesion and hierarchies. Old Bill might realize that Old Bob can whip him any day of the week, but when Old Bob wanders off and Old Charlie wanders in,

Bill hasn't a clue. *Who's this? If he thinks he's getting near any cows while I'm around, he's got another think coming.*

Because of these shifting standings, bulls that wish to procreate during the relatively brief rut must challenge and fight more often than most ruminants. After losing one fight, a bull might rest a couple of days, regain his strength, and successfully challenge a weaker bull with a cow at his side or even the bull that pushed him out of the mating game two days earlier. Combatants can afford to spend their energy budgets wantonly and make up the shortfall later, for bison rut in high summer when grasses are lush and temperatures high. The average bull will lose 223 pounds during the rutting frenzy. Some drop nearly 400 pounds. That's the equivalent of two southern whitetail bucks, a considerable drain on any bull, even the massive giants that can push 3,000 pounds before the rut begins. The prize is worth the cost, however. The more often bulls

The Rut

THE RUT

challenge one another and fight for cows, the more genes they'll pass to the next generation. A strong bull may live to breed for eight years, which means he'll easily produce more offspring than a cow that calves 10 times during her 20-year life.

One study at a modern bison refuge revealed that 23 percent of bulls over the age of four had had their ribs broken at least once. In one dominance fight within a small captive herd, an old bull pinned a young challenger to the ground, his horn buried deep in his antagonist's side, and didn't let up until his rival stopped breathing. Clearly, deciding breeding rights is serious business among bison males. They undermine the long-held belief that rumi-

The bison rut begins in June, when small herds of cows join together. Bachelor bands of bulls will seek these large herds out.

nants act altruistically and rarely try to kill one another during rut fights. Watch the brutal, fierce combat of two bull bison and it becomes obvious that each will willingly and happily skewer the other if just given the chance. But when one hooks, the other counters. When one pushes, the other shoves back. And when one senses it is time to withdraw, he does so at full speed to avoid the charge of the winner. No altruism is evident in such fighting, and why should it be? What does a male lose if he destroys his rival? He rids himself of competition that season and all seasons to come. Never again will he have to risk his own health to fight that animal. Never again will he lose a cow to him. The winner who kills an opponent has increased his own chances for genetic survival. The only reason serious injury isn't even more common during

Moving through a herd of cows, a bull bison lip curls to check a cow for estrus.

bison rutting battles is because each fighter defends himself so well.

The annual bison rut begins in June when small herds of cows coalesce into larger herds. At the same time, small bands of bachelor bulls, usually two to five but sometimes as many as 20 in a loose assembly, break up, its individual members drawn to the growing cow herds like water to a sponge. The resultant mixing may be nature's way of killing three birds with one stone. First, mixing reduces the potential for inbreeding. The more cows and bulls in one place, the better the odds that related animals will not mate. Second, the great amalgam increases competition among bulls, pushing the cream to the top and forcing inferior genes from the breeding pool. Third, having everyone together at one big party assures that no cow will be missed during her relatively short estrus period. There is thus no need for female bison to advertise them-

selves, no need for bulls to defend territories, to collect harems, scent mark, or call out for romance. Just grab your partner and promenade.

To maximize genetic output, bulls must find estrus cows quickly. They do this by moving through the herd, sniffing and testing one after another, lip curling whenever one urinates. During this sampling stage, all bulls are free to join the festivities as long as they stay out of one another's way. As their urge and frustration rise, they indulge in what many zoologists consider classic displacement activity. Since they can't yet breed, they paw, wallow, and horn saplings, breaking off sizable trees and de-barking others. Another theory holds that these three activities, often done together, are vestigial territorial marking behavior no longer needed now that bison are open-country herding animals. The fact that bulls sometimes urinate before they wallow and rub their foreheads in the damp earth supports this idea. Whatever the reason, wallowing,

THE RUT

horning, and pawing are common activities early in the rut.

Because bulls normally do not associate with cow groups, it is postulated that their presence, odor, bellowing, and testing encourage an early and uniform estrus throughout the herd. Whether this is true is as yet unproved, but it works for other species, so why not? Photoperiodicity undoubtedly plays the bigger role in initiating the cows' annual cycle, but bulls cannot indefinitely burn calories and risk injuries while rutting. Shortening the entire program is to their advantage.

A bull pairs off with a cow shortly before her estrus period. The bull is lip curling to check for estrus. Note the open wounds on the cow's side.

As females approach their two-day heat period, they pair off with bulls and circulate through the herd in a classic tending bond. The bull keeps within five feet of his cow, threatening, roaring, and, if necessary, driving off all other bulls, sometimes including the cow's current calf. If the couple is resting, the cow will recline, but the bull will remain standing, on guard. Males that approach too closely are warned away with a broadside present-threat pose, the bull standing stiffly, emphasizing his hump and overall size. Moving through the herd may be the cow's way of influencing mate choice. The more big boys she stimulates, the more they compete for her, and the greater the chance she'll end

up with the best of the bunch.

While preparing his chosen for the upcoming copulation, a bull bumps her with his body, rests his chin on her back and rump, and occasionally grooms her with his tongue. Cows not yet ready for such intimacy may horn their patient Romeo, kick him with a hind hoof, or run away. The bull does not wish to encourage such running, for it excites the herd and draws attention from other bulls, necessitating additional threats, chases, and fights.

Studies at the National Bison Range in Montana have shown that bulls guarding a cow are more difficult for challengers to intimidate than bulls without cows. For instance, Bull 2 might back down nearly every time when threatened at a choice grazing site by Bull 1, but when the same bully tries to frighten Bull 2 from a cow, Bull 2 resists much more often. At that point Bull 1 has two options. He can shrug and find his own girlfriend, or he can escalate from bluff to attack. Interestingly, when he does choose to fight, his chances of winning are only 50 percent. When he only bluffs, his odds for making Bull 2 surrender the cow are much higher. This suggests that bulls do rec-

ognize one another's potential for mayhem. In other words, bulls do not challenge tending bulls unless they think they can defeat them.

This same study revealed a complex social pecking order among bulls that researchers didn't even consider a hierarchy because it changed so often, depending on conditions. A bull's status changed from situation to situation and from day to day, so there was no strict dominance hierarchy as is found among sheep. There was, however, a relative social standing, and bulls with higher social standings did enjoy more breeding opportunities than bulls with lower social standing.

When a bull prepares to mount a cow, he pants to give her warning. If she is ready, she will stand while he raises

BISON RUT: FACTS ON FILE

Family:	Bovidae
Genus:	Bison
Species:	bison
Weight:	Males 2,000 to 3,000 lbs., females 800 to 1,000 pounds.
Height:	Males 6 feet, females 5 feet.
Horns:	18 to 23 inches long, 14 to 18 inches circumference at the base, 18- to 27-inch tip-to-tip spread.
Rut:	July through mid-September.
Habitat:	Grasslands, plains, prairies, mountain meadows.
Range:	In scattered herds from Alaska to Oklahoma.
Society:	Cow/calf herds in fall/winter. Bachelor bull bands. Mixed large

THE RUT

his great bulk over her rump and locks his forelegs around her flanks in a classic clasp. Copulation lasts about 10 seconds, just long enough to plant the seed and attract the attention of neighboring bulls who suddenly rush to the pair, harass them, and follow them wherever they go in the manner of sheep. Unnerved, the tending male bellows warnings and breaks off to chase his tormentors, but not too far or he risks losing his cow in the crowd. That is no

The bulk of bison breeding is accomplished by mid-August, though any cows that have not conceived by that time will have a second estrus cyle in mid-September.
Facing page — The reason for all the rituals and fighting — the next generation.

great loss, however, for female bison rarely allow more than one copulation. When the male determines she is no longer receptive, he begins his search for another likely candidate. Any cows that fail to conceive during their first estrus recycle in about three weeks, so some activity extends into mid-September. The bulk of breeding, however, is finished by mid-August and is done early or late in the day before the great summer heat descends over the prairies. Some eager or insatiable bulls carry on into the night, roaring and pacing, knowing they must strike while they are hot. Soon enough the rut will end, and by next summer they might not be around to join the fun.

Moose

If I say you have a beautiful body, will you hold it against me?

A cow moose is the Mae West of ruminants. No shy, demure debutante, this femme fatale. While elk cows and mule deer does feign disinterest and let the males do the courting, the long-legged moose grabs her beau by the ear with a rousing moan and rubs up against him just to make sure there is no mistaking her intentions. Hang on, Bullwinkle!

Calling makes sense for moose. Much of their north country and Rocky Mountain habitat is densely wooded, and the animals themselves, due to their size and solitary habits, are widely scattered. Under those conditions, it can be difficult finding a Saturday night date. So mamma moose isn't proud. She actively recruits potential mates even as they cruise the swamps and forests grunting and croaking for her. Once she meets the right hunk, she stays near him and aggressively defends him against interloping females, rare behavior among deer.

Like whitetails, moose are largely loners and hiders. Instead of trying to outrun predators, they make themselves scarce — at least as scarce as a seven-foot-tall, 800- to 1,400-pound deer can make itself. Apparently they get the job done,

because they're thriving and increasing in number across most of their range, which takes in all of North America from Alaska's North Slope down through the north woods of the Great Lakes states, the northeast, and as far south as Colorado in the Rockies. They do best in habitat where young shrubs sprout in the wake of fires, floods, and logging, producing an abundance of nutritious leaves, twigs, and buds.

Twins are the norm for hider species, and the moose is no exception. After her first calf, a healthy female in good habitat will often double clutch, and triplets are possible. Though a moose calf isn't nearly as helpless as a whitetail fawn, it hides between feedings for several days after birth. A whitetail fawn will stay in hiding for nearly two weeks. Mamma moose does not wander far for the first two or three weeks of her calf's life as it follows her on unsteady legs. Throughout summer she guards it with a ferocity unknown among other deer species. Bulls, on the other hand, are remarkably docile and slow to anger, despite their great bulk. Only in rut do they get irascible — and slightly loony.

Rutting moose have been known to

chase dogs, humans, horses, and humans on horses, in which case the intent may be murderous or amorous. Who knows? They've also stood on railroad tracks, antlers forward, challenging diesel locomotives. They generally lose those encounters. However, they successfully thrash small automobiles from time to time. Their oddest behavior is unrequited overtures to dairy cows. Every few years some dashing and daring young bull rides out of the north woods and sweeps a docile Holstein off her feet. Farmers, knowing full well that these romances seldom work out, try to discourage the liaisons, but shooing and hazing have limited effect. If a female moose doesn't show up to claim her wandering bull, he may hang around the pasture for weeks in

Two weeks before cows are ready to breed, bull moose begin digging pits with their hooves. When satisfied with his pit, the bull relieves himself into it, then bathes in it.

a mental fog, mooning and dreaming his life away. Bovine infatuations, however, are aberrations. In the normal course of events, moose stick with moose, but even then their actions are a little unusual.

Consider the complicated issue of scent urination, for example. About two weeks before cows are ready to breed, bulls begin digging pits with their hooves. One will stand and scratch with one front hoof like an impatient horse, then switch to the other foot, ripping sod and throwing dirt. Then he'll squat over the hole as if to urinate, but he usually doesn't. Instead, he'll return to pawing. Another false squat, as if measuring. More digging. The ellipsoidal pit may stretch two feet to eight feet, be a few inches or a foot deep, and even have a six-inch berm. When finally satisfied with his creation, the bull relieves himself copiously, then hops in for a good wallow. He may engineer a half dozen such

baths in an evening.

Okay, so the crusty old geezer wants to stink himself up. Most deer do that for the rut. Nothing unusual about that. But here comes a female, and she tries to get into the tub, too, sometimes when it's already occupied. The bull doesn't think much of that, so he runs her off. Wait a minute, a male driving off a female? Wasn't the whole idea of this perfuming business to attract the girls? Well, hang on. Done with his roll, the freshly anointed bull strides out and parades his rank rank, inviting the cows to sidle up and rub a dose of it onto themselves. They oblige and everybody is happy. Everybody except the lesser bulls watching from the wings.

For years naturalists didn't know what to make of this behavior. What exactly

Researchers believe that by rubbing against urine-coated males, moose cows are induced to ovulate.

was going on? In the early 1980s Dale Miquelle of the University of Idaho went to Denali National Park in Alaska to find out. He and his assistants recorded moose behavior around urine wallows for three seasons between August 25 and October 10. They learned that only mature bulls scent urinated in this manner; that young bull urine did not have the strong odor of mature bull urine; that young bulls pirated the smell by rolling in older bulls' pits when they could; that old bulls tried to prevent young bulls from using the pits; that females were attracted to the pits but were chased away by the bull until he was finished rolling; that females competed with one another for the pits; that females rubbed against urine-impregnated males; and that the entire process was done one to two weeks before cows were ready to breed.

Miquelle's interpretation of the evidence? Scent urination is designed to

induce ovulation in females. Makes perfect sense. Like all male deer, bull moose are ready to breed long before cows. This is necessary to take care of the small percentage of females that cycle early. But competing for dominance for several weeks while awaiting the main rush of estrus cows is risky. Bulls can injure or kill one another or simply wear themselves out before their services are required. It makes sense then to hurry the cows along, prime them with an olfactory signal. Within a week or two they ovulate, the job gets done, and bulls can relax and eat again.

Why would young bulls roll in a master bull's pit? Because they don't stand a fighting chance against the much heavier, more experienced old bulls, but, like preteen boys stealing a slap of dad's after shave, they will risk sneaking into a pit for a quick roll when no one is looking. Who knows, bull or boy might wander off and fool an unsuspecting female. Cow moose tend to stay with a mature bull once they've smelled him, so the young pretender, even without sizable antlers, might just pull it off. His initial stolen scent would prime the female, she'd ovulate in a week or two, and bingo. Junior gets lucky. Odds are, though, a mature bull will come along in the meantime to show the cow her error in judgment.

Fascinating as scent urination is, humans are more impressed with moose antlers, which also play a significant role in mating. The obvious question is why such huge, palmated antlers? Once more, the ghost of *Megalocerus* rises. (See caribou chapter.) The antlers of both species look remarkably similar. Yet our moose is not cursorial, is not a runner like *Megalocerus* is thought to have been and

like caribou are. As already described, moose calves are hiders, not herd followers. They don't have to be born ready to run. So why do moose grow so much antler? Actually, they don't. It only looks like a lot of antler. In relation to body size, moose antler mass is about equal to that of mule deer and elk. Nevertheless, it is still the largest antler in the modern world, and it illustrates a curious natural phenomenon.

Common sense would suggest that animals living in the north where nutrients are locked under snow and ice for nearly half the year would husband their energy resources. Species in the tropics, with their year-round growing season, should be the ones to splurge on accessories like huge antlers. Yet tropical deer have tiny antlers, mere spikes in most species. What's going on?

Despite balmy temperatures, or because of them, tropical habitats are relatively sterile. Organic decay is so rapid that topsoil has no time to build. Nutrients like leaves are recycled almost as fast as they fall. At the same time, competition for resources is fierce, which forces plants to evolve defenses such as toxicity and thorns to keep from being devoured by insects and animals of various stripes. In such environments, mammals are forced to minimize body size and fiercely defend territories and the limited forage within them. Tropical deer are thus small, few in number, and limited in species diversity. They grow spike

Moose antlers are the largest grown by any deer. Their canoe paddle shape and incredible mass are used to fight and intimidate other males and to attract females.
Next page — Bull moose shove each other to establish dominance.

122

antlers because that's all they can afford and because spikes are most effective in punishing trespassers. Forget ritualized fighting. It's a jungle out there.

Farther north, cold winters temporarily stop bacterial action, essentially banking the summer's sun energy in dead organic material that builds rich, productive soils over time. This stored energy bursts into flower suddenly and profusely. Young plants are rich in vitamins, trace minerals, and protein, and low in toxins. There is more than enough to go around, so herbivores do not have to defend any of it. They can live and forage side by side, develop intricate social

Two old, weathered combatants give each other the evil eye.

behaviors, large bodies, and something zoologists call "luxury organs." Showy body hair like manes are luxury organs. So is fat. (You don't find much fat on tropical mammals.) The moose's bell, that pendant dewlap of flesh and hair hanging from its neck, is a luxury organ, its purpose unknown. It is probably just a swinging visual signal to attract attention to the head and antlers. And the ultimate luxury organ is the antler. In a sense, northern deer grow lavish antlers because they can. Why not? There's all that energy out there, all those nutrients. What do you do with it all? Might as well put it to work. Might as well grow a rack to make the girls swoon and the guys run for cover. What have you got to lose?

Over the millennia as glaciers waxed and waned, dozens of antlered species arose and died out in the north, evolving ever larger and more extravagant antlers, using them to fight, to intimidate, and to curry favor with females. As mentioned in the chapter on antlers, Alaskan moose are the current record holder for growing the most antler in the least amount of time, manufacturing 90 pounds of bone in as little as 90 days. During the peak June growth period, they pile on about half that total, a phenomenal rate of bone tissue formation. Southern moose, like the smaller Shiras subspecies of the Rocky Mountains, enjoy a longer growing season and more measured antler growth.

Contributing to the impressive appearance of these racks is their unusual, canoe paddle shape, which seems designed for triple duty. First, like most antlers, moose palms signal genetic health and vigor to prospecting females. They can't miss those broad slabs of white bone. Second, held low and twisted like semaphores, they serve as warnings to rival males. And third, the palms, which can measure 4-1/2 feet long and 2-1/2 feet wide, function as shields to ward off enemy tines

during moose shoving contests. A typical moose match will see one bull treading backward under his opponent's onslaught until he suddenly bucks up, braces his legs, and shoves back, reversing the tide. The bull that gains the greatest momentum can literally push his adversary down or, more often, turn him, at which point he presses the attack into the loser's side or haunches. Bulls that do not spin and escape quickly enough can take a significant beating. During a study of moose hides, Valerius Geist found 225

MOOSE RUT: FACTS ON FILE

Family:	Cervidae
Genus:	Alces
Species:	alces
Weight:	Males 1,000-1,400 lbs., females 800-1,000 lbs.
Height:	6 to 7 1/2 feet, depending on subspecies.
Antlers:	Bases 8-12 inches around, spreads over 6 feet, palms as wide as 2 1/2 feet, as long as 4 1/2 feet, as many as 39 tines. Weigh up to 90 lbs.
Rut:	Early September through mid-November.
Habitat:	Conifer forests, taiga forests, tundra, second-growth woods, swamps, bogs.
Range:	North slope Alaska east to Atlantic coast, south to New York, Great Lakes states, Rocky Mountains south to Colorado.
Society:	Cow/calf family units, lone bulls and bachelor groups except during rut. Bulls winter alone, cows with

antler wounds on a 10 1/2-year-old bull, 199 on a mere stripling 3 1/2-year-old. Other studies have shown a death rate from fighting somewhere between 4 percent and 10 percent annually among all deer, not just moose.

Normally, of course, intimidation and ritual play the biggest role in bull-to-bull conflict. As with all deer, moose males spend the summer alone or in twos and threes with little social interaction, piling on fat and antler bone until the waning light of late summer gets their testosterone and their danders up. By early September they are ravaging the willows, stripping velvet, and shadow-boxing for the upcoming fight season. Some start so early that they break off tine tips that haven't had time to fully mineralize. Predictably, youngsters spar regularly and eagerly. Old timers save their energy for

A bull checks a cow for estrus. His big, overhanging nose prevents a classic lip "curl."

the real thing, which rarely occurs until late September when there is a cow moose to win.

Like all intelligent fighters, prime bulls measure their opponents carefully before wading in. Then, whether they think they can whip them or not, they try to "talk" them out of the whole idea with croaks and groans and sign language, beginning with a stiff-legged, swaggering dominance display. Head up, ears back, mane erect, each bull trots around the other, tipping his antlers side to side with each step, flashing the big paddles in warning. This is sufficient to make a significantly smaller adversary leave the arena, but closely matched foes require more convincing. This could be a vigorous round of thrashing brush and saplings during which sizable spruce trees are demolished. If neither bull feels sufficiently intimidated to leave nor antagonized to attack, both may begin displace-

ment feeding with exaggerated intensity, jerking their heads up, ripping vegetation viciously, and staring at one another. This can drag on for hours, sometimes days, until one wearies of the game and departs or a cow reaches estrus and forces the issue. Then the fight is joined as mentioned, a great pushing match. Rarely antlers spring apart, then snap back together in a death lock, dooming both males. Wolves and grizzlies soon find them. Clearly, intimidation is the better option.

As a cow nears her heat, bulls stick close, acting out stereotypical deer rut behavior. They approach cows in a low stretch, sniff their urine and lip curl, though lip "lift" would be a more accurate description. Their big, overhanging noses preclude any serious curling. They rush smaller males that drift too close, and thrash brush to keep the competition guessing. Urine wallowing nearly ceases once cows enter estrus, more evidence that the enterprise is designed to induce ovulation rather than declare dominance.

Although bulls do not defend harems, in open tundra and taiga habitats such as Denali National Park, as many as a dozen cows may gather around a master bull and compete with one another for his attentions. This aggregation attracts other bulls and increases competition, and is possibly a tactic to sort out the prime mates. Naturally, cows won't settle for second best once they've identified the real thing, which they probably do by assessing antler size, body size, and dominance behavior.

Surprisingly, attitude and experience can sometimes win the day for an older bull who goes up against a younger male with larger antlers. With a combination of bluff and aggression, the old timer can back the youngster down and win, but it is doubtful he can hold out for the entire two-week period when 89 percent of all cows are bred. As a general rule, the biggest bulls are the early breeders, and as they wear out, the next in line take over.

Each cow remains ripe for one or two days and will stand for several mountings in that time. It is not known if there is more than one copulation. The bulk of breeding occurs between the last week of September and the first week of October in most regions, but late-cycling cows extend breeding into November. Like all female deer, if a moose cow isn't impregnated during her first heat, she'll recycle and try again. This is when the younger bulls get their chance, for the superior bulls that controlled the early rut will already have withdrawn to recuperate. Their testosterone suddenly depleted, many shed their crowning glory by late November. Most are antlerless by January. Only youngsters keep their tiny racks into March. By then the old timers will be showing velvet buds. Winter isn't even over and already they're preparing for the next big dance.

Mule Deer

When it comes to love I want slow hands

If the mule deer were a bird it would be a duck — an odd duck. For a deer it's rather blocky and heavy looking, especially old bucks during the rut. They look like pot-bellied old truck drivers. Rather than running and leaping gracefully when alarmed, a mule deer bounces like a kid on a pogo stick. And those large ears — well, they don't call it "mule" deer because it's stubborn. The animal has a reputation for stupidity, too. Where a whitetail will flee in terror, a muley will stand and stare. And even when it does catch a clue and hasten away, it will often stop and look back, sometimes even prance back as if needing a pinch to believe what it's seeing.

Such behavior does sound dimwitted and ill-fitted — until you consider one salient fact: this is the only North American Cervid that has pioneered, populated, and thrived in our western sagebrush flats, desolate shortgrass plains, high-mountain conifer forests, and deserts. The highly respected, adaptable whitetail has not been able to do that, nor has the elk. So there. Our big-eared deer isn't odd, it's unique — designed for unusual conditions in unusual places. What appears to be stupidity, slowness, or indecisiveness is actually intelligent caution and perhaps a healthy dose of genet-ically programmed equanimity that just might be selected for in the mating tactics of this species.

The biggest reason the mule deer is so unusual is because it's the new kid on the block, our most recent species, a native American through and through. While elk, bison, mountain goats, and sheep all wandered here from Asia, the muley evolved from the original line of New World deer — whitetails, which themselves arose on this continent from an ancient line some 4 million to 6 million years ago during the Upper Pleistocene epoch. Sometime in that dim past white-tails colonized much of North America, coast to coast.

There are at least three schools of thought as to how the mule deer emerged from the whitetail line. One has it that whitetails adapted to the cold, dry West and became the mule deer, which in turn moved west and north to the Pacific coast to become the Columbian blacktail and Sitka blacktail. A second theory reverses that, making the Sitka of Alaska and British Columbia rain forests the direct descendant of whitetails and thus the old geezer of the blacktail line. That makes sense to casual observers because Sitkas look like whitetails. Compare antlers from each and you couldn't tell the difference, except the Sitka rack is much smaller and doesn't have brow tines. Sitka tails are fairly wide and bushy, too, like the whitetail's, and Sitkas have almost no rump patch. Mule deer have big, white rump patches, skinny little tails, and tall antler beams that fork (bifurcate) and fork again. Sitkas do have distinct black forehead "shields" reminiscent of a mule deer's facial mask.

Supporting this north-to-south evolutionary progression is what biologists call a "cline" of mule deer subspecies progress-ing down the Pacific coast, then turning inland to the deserts and Rocky Mountains. Each subspecies down the line looks more like a classic Rocky Mountain mule deer, with progressively larger bodies, antlers and rump patches but smaller tails. Thus we have the Sitka, Columbian blacktail, California mule deer, Inyo deer (perhaps a hybrid or intergrade between the California and Rocky Mountain mule deer), and finally the King of the Rockies himself, Mr. Mule Deer. Larger body size, antlers, and rump patches are marks of advanced evolutionary species, so this second theory has strong backing.

The third theory, postulated only recently by Dr. Valerius Geist, professor of environmental sciences at the

The mule deer is considered a relatively new species, having evolved from whitetails.
Next page — Mule deer do not keep harems. They guard a single female which is near her heat cycle, and the rest of the does just happen to be there because mule deer females live in small groups.

University of Calgary, Alberta, and a world authority on mule deer, is stunning in its novelty. Based on DNA analyses by Mathew Cronin of Yale University, Geist theorizes that the Columbian blacktail did indeed evolve from whitetails, but then those two species later got back together to produce a hybridized mule deer. Here's what Geist thinks happened. After west coast whitetails had been separated from the rest of the continent's whitetails by climate, topography, or habitat long enough to evolve into blacktails, climatic conditions arose that permitted each species to pioneer toward the Rockies. Whitetails went west, blacktails went east, and where the twain met blacktail bucks frolicked with whitetail does and produced mule deer fawns. At least that's what the DNA findings suggest. In case you've misplaced your high school biology text, DNA is that building block in our cells (deoxyribonucleic acid) containing the genetic code. Mitochondrial DNA (mtDNA) is passed on only by females, which means maternity can be traced back thousands of generations, virtually forever. Cronin's tests revealed that the mtDNA of mule deer was nearly identical to that of whitetails. Whitetail and blacktail mtDNA was only similar.

This hybridization wasn't the result of a one-night stand, mind you. Whiteails and blacktails bred many times over many years, so often that the hybrid offspring began breeding to create a pure line of mule deer. When changing climate and habitat (continental glaciers, southwest deserts, the Great Plains) again pushed whitetails and blacktails far apart, the newly created mule deer was able to thrive in the dry, rugged land between.

This is mostly theory, of course, based on as much hard evidence as science has been able to generate. Thanks to repeated glaciation, most of the fossil record has been wiped from the geologic slate over which prehistoric whitetails and blacktails evolved. The DNA record, however, proves incontrovertibly that mule deer are closely related to whitetails through the maternal line. Supporting the notion that muleys are newly evolved is the fact that, though they have spread from central Mexico north to the Yukon, they haven't yet changed physically to meet those climatic extremes. Northern specimens suffer frozen ears, for instance. Caribou know better (genetically) than to grow large ears, but then they've been evolving in the arctic for millions of years. Given time, natural selection may produce a northern mule deer with shorter appendages and increased body size (conserves body heat), a southern variety with a smaller body and larger appendages (disperses body heat) as predicted by two old axioms of biology, Bergmaus' and Allen's rules. Whitetails have already made these regional adaptations as seen in the tiny, 50-pound Florida Keys whitetail and the 300-pound Dakota whitetail.

Because mule deer are a new species, it isn't surprising they exhibit a strange mix of rutting behaviors, some similar to those used by whitetails, but others suggestive of elk. For decades casual observers and more than a few experienced naturalists reported that mule deer bucks controlled harems. Visit mule deer country in mid-November and sure enough, there stand five or 10 does with a big buck in attendance and perhaps two or three smaller bucks flirting on the fringes. If one of the youngsters gets too close, the master buck cuts him off, some-

times rushes him, and drives him at antler point from the area, just as a bull elk will drive off its competitors. A master muley buck seldom has to drive off another large buck because those guys will be some distance away with their own cluster of does. Just like elk. Nevertheless, these muleys are not guarding harems. Rather, they are protecting a single female, the one nearest her heat cycle, just like whitetail bucks do (serial polygyny.) The rest of the does just happen to be there because muley does live in groups, probably matriarchal clans. Whether they want them or not, bucks get a whole family of chaperones with their date. A whitetail buck takes his doe for a wild run through the countryside, then the pair finds a quiet corner and mates. The muley stands calmly beside his

Mule deer bucks don't scent mark or defend territories like whitetails because they often migrate many miles from summer to winter ranges to reach suitable feeding sites.

doe while her mother, aunts, and sister look on.

This "part whitetail, part elk" behavior is a result of where mule deer live — semi-open country. One day they might be ensconced in a heavy stand of conifers, the next they could be feeding on a broad grassland. Some live nearly year-round in heavy forests, others are at home on the plains where they're lucky to find grass higher than their knees. In open habitats they may be developing stronger herding tendencies, like elk, for predator protection. Mating behavior could be evolving to match. If master bucks, while guarding single does within a group, inadvertently reserve the entire group for their exclusive breeding, natural selection might eventually convert the species into harem breeders.

Migration is another product of open-country living in temperate climates where forage is here today, gone next

December. In the mountains muleys migrate down from summer to winter ranges and sometimes move many miles in flatter terrain to reach suitable winter feeding sites. For this reason, males do not defend territories, and this, too, is reflected in their mating behavior. Whitetail bucks mark their home grounds by rubbing scent on tree limbs, then scraping a bare spot on the ground into which they urinate, dribbling the urine over their tarsal glands. If a doe stumbles onto one of these scrapes — and she can hardly miss them, they are so obvious — she'll trickle into it a "come hither" message before moving on. Like a Casanova flipping through his black book on Friday night, a buck will check his scrapes until he detects hot doe scent. Then he follows

Young mule deer bucks often "buddy up" to a large buck to take advantage of his status to bully other males. This social paratism is unknown among other deer.

his nose.

If a muley tried that, he might die of loneliness because all the eligible bachelorettes had moved down to the valley alfalfa fields. So instead of scent-dousing the ground, a muley buck stands over his hocks, presses the erect hairs of his tarsal glands together, and urinates on them. They soak up the stink like an elk's belly hairs and the buck becomes a mobile olfactory factory, carrying his perfumed invitation with him. In effect, he becomes his territory, spreading the news of his virility wherever the wind blows. Rutting bucks and does predictably circle downwind of larger bucks to sniff their tarsal glands. Apparently a master buck smells more powerful or commanding than a subordinate buck. This reduces the need to fight to determine who's top dog.

Antlers, of course, are the classic status symbols among deer, and muleys wear some beauties. Beams can grow to seven

inches in circumference, 30 inches long, and spread 35 inches wide. That's enough to intimidate any would-be challenger or impress any reluctant doe. Research has repeatedly shown that bucks recognize one another by their antlers, which also prevents a lot of needless fighting. However, the largest-antlered buck doesn't always sit at the top of the totem pole. In stable populations where deer know one another well over several years, a massive, old buck with regressed antlers can dominate younger, more heavily-antlered competitors. Body size seems about as important as antler size, but the two usually go hand in hand. Through belligerence and bravado, a slightly smaller buck can often dominate a meeker but larger one.

In keeping with their eccentric reputations, mule deer practice a social parasitism unknown among any other deer. Like an insecure, scrawny human, a young buck will "buddy up" with a master buck early in the rut and borrow his status to bully other males. This peculiar relationship begins innocently enough with a sparring match. A dominant buck will solicit a subdominant male, usually one or two age classes younger, by repeatedly feeding near him and occasionally displaying weak appeasement postures such as slightly lowering his back or crouching. As the smaller animal becomes assured that he isn't about to be trounced, he begins to relax and trust the larger buck. Eventually the big guy will maneuver his way in front of or beside his would-be partner and turn his lowered head to him, offering his antlers. Once the subdominant gets the idea, he touches antlers with the dominant, and a tentative wrestling match begins.

In the initial stages of this sparring

Sparring among young mule deer bucks is extremely common, both before and after the rut.

Two evenly matched bucks face off. Body hair is erect, ears laid back, tarsal hairs flared ... note that the buck at left has its tail tight against its butt, a sign that it is indecisive and will likely back down from this confrontation

contest, the smaller buck jumps back frequently, as if still unsure of the arrangement, but his adversary waits patiently, avoiding eye contact. Reassured, the lesser buck dips his antlers and rejoins the sparring. Thus, in fits and stops, twisting and pushing, the match escalates for many minutes to an hour, each participant becoming bolder and more informal. Eventually the dominant may be able to initiate antler contact without frightening the subdominant. The pair may then forage, spar, and bed together for several days as buddies. Throughout the relationship the dominant buck urinates on his tarsal glands while his young friend assumes the squatting, subdominant urination posture of a doe.

Should another dominant buck intrude into this cozy friendship, the small buck may brazenly confront him with a flagrant dominance display, something he wouldn't dare try alone. But with his big buddy there to back him up, he can often put the larger stranger to flight. Of course, if the little toady presses the chase too far beyond the influence of his protector, his victim may suddenly swap ends and run the arrogant little twit back with his tail between his legs.

Dr. Geist suggests the payoff for this unusual relationship may be a symbiotic breeding behavior latter in the rut. A master buck will tolerate his former sparring partner quite near the doe he is courting. Geist theorizes that the master buck may benefit because his partner chases off smaller bucks, thus freeing him to concentrate on breeding until a buck his own size shows up. His minion reaps an occasional mating of his own while the boss is busy routing a serious challenger or when too many does come into estrus simultaneously.

Such relationships are not common, but sparring is, especially among young deer before and after the rut. During the 1975-76 desert mule deer rut in Big Bend National Park, Texas, Thomas Kucera observed 86 sparring matches, 71 percent of them involving young spikes and forkhorns. More than half of all contests were between bucks one antler class apart. Only 30 percent were between bucks of equal antler size. Kucera noted that during breaks in sparring, participants lifted and turned their heads broadside to one another as if displaying. This could be interpreted to mean the fighters are "sizing" each other up, learning how antler size and possibly neck and body size relate to strength. Knowing such things could reduce unnecessary fighting later. That could explain why older bucks don't spar much. They already know about size/strength relationships. Or it could be that young bucks, like young boys, just like to wrestle.

There is no mistaking a bit of friendly tussling for a real fight. Muley bucks with a point to prove go at it hoof-to-hoof like a couple of half-drunk loggers, snorting and drooling and throwing their substantial weight around in a sincere effort to kill their opponent. And a muley fight is usually over a doe. Because of their violence and potential for destruction, such fights aren't common. Most are avoided through social ranking and dominance displays. A dominant buck advertises his status with sight, scent, and sound. First, he looks big and tough. In preparing for the rut, a five- to eight-year-old in his prime will pile on fat and grow a

grotesquely thick neck. Such a beast can weigh 350 pounds. In early November (December in some southern ranges) this hunk of muscle and bone will saunter down from his summer range high in the mountains (unless he's a desert or plains muley) looking for love. He will stop frequently to "horn" bushes and saplings, either violently "fighting" and breaking branches or more gently rubbing his antler bases and forehead on them. These horning sessions can last 15 minutes and appear to be used to burn off excess energy, psyche-up the buck doing the horning, and to challenge other bucks. When a buck stops horning he will stand quietly and listen. If he hears another buck horning, he lays his ears back, increases his apparent size by erecting his body hairs (piloerection), and trots toward his competition.

Next comes a serious session of bluff. If two bucks are nearly equal in body and antler size, they approach one another at an angle in full display. Body hair is erect, ears are laid back, tarsal hairs are flared, preorbital glands open, tail raised at a 45 to 90 degree angle from perpendicular. (An indecisive buck holds his tail tight against his butt — a sign he'll lose the confrontation.) They walk stiffly, often stopping to violently horn a shrub or rub-urinate. If neither backs down, they stiffly walk closer, then circle one another. Now one may try his final bluff, the grunt-snort, a sudden gutteral grunt combined with a sharp blast of air through the nose. It's loud and so surprising that it often makes the other buck flinch. If he runs, the display worked and the fight is prevented. The dominant buck chases the loser and in effect yells "chicken" by grunting and pounding his front hooves on the ground.

If, however, the grunt-snort is ignored, it's time to rumble! From a body length apart the duelers launch into one another, clashing horns. At this point one may turn and run if he realizes he's outclassed. Otherwise they begin pushing, twisting, pulling and grunting in an all-out brawl. Each keeps his center of gravity low on crouched, wide-spread legs. The front legs are splayed forward to prevent being pulled in that direction. Victors may bowl their opponents over, flip them onto their sides, and even toss them over their backs. Tines may break, and entire antlers may snap off. These fights are incredibly fast and furious. Usually the first buck knocked off his feet loses. If he can't parry his antagonist's thrusts, he must flee or suffer serious gouging, possible impaling and death. Torn ears, ripped eye lids, and gashed and tine-poked necks are common battle scars.

This violence and its speed are all the more extraordinary in light of the mule deer's calm temperament and cautious nature, evident in the way it evades predators and courts females. When a mule deer sees or hears a potential enemy, it does not panic. Instead, it calmly walks away or stays put and keeps track of the disturbance. This is what humans misinterpret as stupidity. If a buck thinks it can successfully hide, it lies down, even stretches its neck on the ground. A cautious muley may remain in this position until nearly stepped on. As a result, many predators pass by, never the wiser. After they're out of sight, the deer sneaks off in the opposite direction. If discovered, the deer blasts out in its peculiar bouncing gait called stotting. Stotting allows a muley to ricochet in any direction including backward and to efficiently ascend

MULE DEER RUT: FACTS ON FILE

Family:	Cervidae
Genus:	Odocoileus
Species:	hemionus
Weight:	Males 175-350 lbs, females 150-250 lbs.
Antlers:	Bifurcated, beams to 30 inches, spreads to 35 inches.
Rut:	Late October to early December in north. A month later in southern ranges.
Habitat:	Conifer forests, sagebrush & grassland foothills, alpine mountain tops, deserts, shortgrass plains, brushy canyons, oak woodlands, sage brush flats.
Range:	From central Mexico north nearly to Alaska. From the Dakotas west to the Cascades and Sierras.
Society:	Female clan groups, small herds, large wintering herds. Summer bucks in bachelor bands, winter bucks alone until antler drop, then join female herds.

steep terrain where runners have trouble following. That's why these deer are so fond of canyons, badlands, shrublands and similar broken terrain where they can use stotting to put obstructions between themselves and attackers. Muleys almost always escape by stotting uphill. Whitetails almost always escape by running downhill.

A group of does and fawns breaks from a thicket like a covey of quail. Noise and confusion. Rump hairs flared, feet pounding the ground, each zigzags in different directions, distracting pursuers and warning nearby deer. If not pressed, such escaping deer will stop to try to locate their attacker or reorganize as a group rather than leap blindly. They may even prance back, heads up, ears turning, and noses flaring to pinpoint danger. This may be a genetic response to hundreds of thousands of years of living and dying with pack-hunting wolves that spread out and flank their prey. A muley can stott in any direction to evade a predator's leap, but only if it can see the coming attack. Only after ascertaining a clear escape route do muleys line up and head out. They'll run miles and completely abandon a drainage where a cougar has made a kill. If pressured repeatedly, even by coyotes, doe clans will break up and scatter to hide in rough terrain or dense woods, possibly a throwback behavior to ancient whitetail genes.

This explains why bucks are so patient and seemingly plodding in their courtship. If a muley doe suddenly took off stotting in response to an aggressively courting buck, she might alarm all her sisters, most of which would follow her in panic. Fawns might become lost and the herd severely disrupted, making it easier

for the big bad wolf to come knocking on the straw house door. Running does would also attract the attention of other bucks. Thus it is in a buck's self-interest to court cautiously. Because he tends one female at a time, a buck's job is to find the one closest to estrus. That means he has to test many over the course of a day, which he does by sampling their urine with the standard lip curl, pushing their sexual pheromones into the Jacobson organ atop his palate. This test, or course, requires a urine sample, which does are as reluctant to give as are some Olympic athletes. Patient bucks, however, bide their time until one doe in the group finally squats. Then he steps over, collects the sample, and tilts his head back to inhale the bouquet with all the flair and concentration of a wine connoisseur. If the urine was served before its time, he abandons that doe and waits for the next sample.

Some males, overcome with anticipation, lose control and regress to the rush-courtship of their ancestral whitetail line. A rush-courting buck glares at a doe, tenses, sometimes squeezes out a long, low blat, then attacks the poor doe, head down and antlers leading. He absolutely loses it, blows his cool. The object of his desire flees in terror with the brute lunging, coughing, grunting, and literally roaring right behind her. But there is method to his madness. Abruptly he stops in a classic low-stretch posture, head and neck parallel to the ground at the level of his shoulders. She stops too, alert, nervous. Then, seeing the danger has passed, she relaxes and relieves herself. Bingo. He walks over and samples the sample.

The job of a mule deer buck is to find the doe closest to estrus. He does this by sampling the urine of many each day.

THE RUT

When a female's estrus is near, the male follows her closely, playing on her maternal instincts, shamelessly pretending he's a fawn.

The contrasting pattern on a buck's face, when seen head-on, creates a foreshortening effect that makes it look like a young fawn. The impostor lowers his head to fawn level, tilts his nose up, looks directly at "mamma" and sounds his best imitation of a fawn bleat.

This blatantly patronizing behavior doesn't fool the doe, but it tugs at her genetic heart strings enough that she often lets him get within touching distance, something most female ruminants won't allow until they are ready to breed. He continues following, flicking his tongue, licking his nose, and softly buzzing. Whenever the doe looks at him he bobs his head up and looks away in

When a mule deer doe's estrus is near, a buck may behave like a fawn to win her favor.

standard mule deer appeasement behavior, as if she's the farthest thing from his mind. Oh, he's a sly devil.

Impatient young bucks, meanwhile, harass any does beyond the immediate influence of a master buck, pushing them until they lie down, seek escape in heavy brush, or rush into the arms of the old master buck, who grins and bears it. Just by being polite, patient, and calm, he gets the benefits of a harem without the work of herding one. And that might be how female mule deer select breeding partners with a genetic predilection for equanimity and caution, just what their fawns will need to survive. Bucks, of course, weed themselves out through the usual competitive processes: living long enough to grow large antlers, escaping predators and disease, eating well enough to produce massive antler growth, laying up enough fat to remain vigorous throughout a month of rutting, plus horning, rub-uri-

nating, sparring, and fighting to establish the dominance hierarchy. This winnowing of the "also rans" takes pressure off females and makes their selection process easier.

When a doe is finally ready to breed, she permits the male to nose and lick her, lay his chin on her back and flank, and mount her numerous times prior to actual copulation. Researchers have counted over 40 precopulatory mounts. Copulation itself involves one thrust, the buck's hind feet often leaving the ground, his head thrown up and back. Quality over quantity. Immediately after copulation, the buck stops his gentle tending behavior. He'll browse, chase other bucks, and investigate other does. If none are ready to tumble, he'll return to the same doe, which might stand for several more breedings during her 24- to 48-hour estrus. But then her chemistry changes, the romance is over, and he knows it. *Ariva dirche* baby. It's up to her to raise the fawns alone.

Copulation among mule deer involves one quick thrust after as many as 40 precopulatory mounts.

Occasionally, a tired old buck may need some encouragement to rekindle his desire, usually late in the rut. At such times a doe will play the aggressor, prancing enticingly, rubbing him where he would normally rub her, even mounting him.

After the hectic 20-day mid-November breeding circus, bucks look like they've stayed too long at a college frat party. Their coats are torn, tines broken, bellies gaunt, ears ripped, eyes glazed. They probably haven't had a decent meal in a month or slept more than a wink. Enough is enough. They quit. Just like that. They stop rub urinating and begin squatting like females. If a doe comes into heat late, they let the young bucks breed her. All they're interested in now is finding a quiet corner to sleep it off. Then they begin the long road toward recovery, foraging heavily, trying to regain body condition before the worst of winter sets in. Many don't make it.

During his research in Canada's Rocky Mountains, Geist discovered an unusual buck rutting strategy. A few large, old bucks opted out of the festivities. They just said no to sex. No chasing, no fighting, no romance. Instead, they kept to themselves, conserved their fat reserves, and consequently had an easy winter. Sound like a recipe for genetic suicide? *Au contraire.* Geist discovered that some of these non-breeders were merely biding their time until competition declined. After a hard winter killed most of the previous season's exhausted master bucks, the fat, healthy, non-breeders emerged to mate virtually uncontested the next fall. Geist figured that some of them fathered as many fawns as their more aggressive predecessors.

Dimwitted mule deer?

Muskox
Ain't no mountain high enough

It would be neither wholly inaccurate nor unkind to describe rutting muskoxen as two haystacks making love. This isn't a creature prone to exhibitionism. Guard hairs as much as two feet long drape nearly to the ground, all but obliterating the plodding, phlegmatic animal beneath. Only a square, black muzzle and two hooked horn tips betray the front end, a great hump the shoulders, and four blunt hooves the bottom. The shoulder hump is so high and the neck mane so thick that the beast's head appears to grow out of its knees. The muskox is at best a slightly animated hairball. But inside that great mound of concealing hair lies nearly half a ton of procreative dynamite that will not let a poorly designed mating form stand in the way of self-perpetuation.

Like bighorn sheep and whitetails, muskoxen are relics of pre-glacial times, associates of mastodons and cave bears. Unlike sheep and whitetails, they look it — slow, plodding, and phlegmatic, with the apparent grace of a boulder, the sort of creature you expect to star with Raquel Welch in a cave man movie. Muskoxen mostly graze, ruminate, and rest, fair weather or foul, riding out Arctic blizzards and -50°F temperatures as if they were born to them, which, of course, they are. These relatives of sheep and bison epitomize survival strategies at extreme latitudes. Their bodies are massive and abbreviated to maximize heat retention, their extremities short to minimize heat loss, their qiviut underhair fine and thick to hoard every BTU they generate. So forget the articulated, graceful movements of long-necked deer or long-legged pronghorns. Forget elaborate courtship displays and subtle strategies. Muskoxen are into straightforward propositions and blunt trauma.

For years the phylogenetic relationship of muskoxen has been in question. Some taxonomists aligned it closely with bison, others with sheep. These days it is considered a close cousin only to the takin *(Burdorcas taxicolor)*, a goat-like Himalayan animal, yet it practices the rutting behavior patterns of several other common ruminants. Muskoxen control harems like elk, roar like bison, butt heads like sheep to settle dominance, and display broadside with heads turned away like mountain goats. Unlike any other North American ruminant, however, muskoxen use their large preorbital glands in a peculiar fashion to threat-display and to scent mark their own legs.

Muskoxen bulls enjoy the relatively easy pleasures of harem rutting thanks not to their own herding efforts, but to

wolves. Because of wolves, muskoxen are herd animals. When threatened, they form their famous defensive circle, calves in the middle, and impale with their sharply hooked horn tips any toothy predator that cares to try its luck. Fewer than five oxen make a poor defense, so each animal instinctively joins with at least four others. Herd sizes commonly increase to fifteen, but then may easily break up because five or more can drift away as a group and still feel secure. Small groups scattered across the tundra reduce in-fighting and grazing pressure, making it easier for each member to grab all the vegetation it needs.

Taking control of such a compact herd is a simple matter when testosterone moves a bull in July or August. He simply joins a group and defends it against other bulls. Policing the cows is almost superfluous because they're afraid to wander off anyway. A bull merely walks toward a female he thinks is getting out of line and she hustles back to the girls. A harem can increase when two cow groups meet and mingle, but a bull's odds of keeping more than 20 under control are poor. Instead of courting and breeding, he begins draining himself walking from one end of the widely grazing bunch to the other. Almost inevitably a splinter group moves off, and he's torn between two lovers, so to speak. Without so much as a grumble of complaint, he chooses one and gets back to business.

In most muskoxen populations, bulls outnumber cows. The males are fertile by age two, but rarely breed until age six. At that age they've reached maximum body and horn size, the two prerequisites for a harem master. The upshot is a goodly number of slighted bachelors wandering the tundra, looking for love in all the wrong places — that is, harems that have already been claimed. When an unattached bull spies a herd, he stakes it out from several hundred yards, watching and sizing up his chances. He may make his move almost immediately or hang around for a day, building up his courage. But he almost always makes a play for it. Nothing ventured, nothing gained.

This attitude isn't surprising, since there is little external indication of a harem bull's fitness. Hair covers a multitude of sins. There are no huge, obvious antlers or horns to indicate relative dominance. If a bull wants to test the waters, he has to take the plunge.

The muskox plunge starts with a stiff-legged walk toward the opponent, head held low, flat face perpendicular to the ground, nose almost touching the tundra. This positions the broad, horny helmet or boss (the broad base of the horn that covers the skull and is the bull's chief weapon) forward like the battering ram it soon will become. This is an obvious threat, and the herd bull usually walks out to intercept it as if knowing he has the psychological advantage because the harem is already his "territory." Before bumping heads over the affair, both antagonists emphatically suggest the other reconsider. They tear up sod with their horn tips and vigorously rub their preorbital glands on the upended sod as well as on their extended lower legs. This gland rubbing has been interpreted to be everything from sharpening the horn tips to grooming the hair to keep it out of eyes during fights. But most behaviorists now understand it to be a threat display, visual and probably olfactory, like schoolboys shaking fists and spitting at one another.

Hair covers many flaws. For a male muskox to challenge a harem master, he has to fight.

It is most often performed by dominant animals within the herd, but is behavior all muskoxen will indulge when confronted by wolves. Placing one foreleg forward suggests approach, lowering the head exposes the boss, and jerking the head up and down to rub the gland tosses the horn tips aggressively.

The next level of threat is the head-tilt and parallel walk, common among hoofed animals. Muskoxen bulls perform this at unusual range, often more than 100 yards apart. Both may march broadside in the same direction or opposite directions, or one might walk while the other watches. Each tilts the top of his head toward the other and turns his face away as if showing the size of his boss. During these preliminaries, either bull may roar like a lion, trying his best to force his competitor to

retreat, but in vain. Researcher Timothy Smith studied rutting muskoxen during 1972 and 1973 on Nunivak Island in Alaska and reported in his master thesis that when a lone bull approached a harem a "dominance battle was almost a certainty." He determined that threat displays might bolster the combatants' courage or intimidate their rivals, which would limit the intensity of the upcoming battle, but not prevent it.

Smith noted that most threat displays lasted more than 10 minutes and could involve several approaches and temporary retreats until the challenger worked up the courage to attack. One bull moved in, displayed, and backed off five times in seven hours before finally screwing up his nerve and attacking.

A muskoxen fight is remarkably similar to a sheep clash. From 30 to 50 yards away, each gladiator takes aim at his nemesis and accelerates to 25 miles per hour, landing horn first with a crack that can be heard for more than a mile. Considering that a prime bull can weigh 900 pounds, this sudden impact is the equal of driving a sub-compact car at 50 mph into a large oak tree. Fortunately, muskoxen have air bags, or, more precisely, something of an air head. Beneath four inches of horn helmet lies a pneumatic skull more than three inches thick. The first vertebrae fits into a notch at the back of the skull, aligning it perfectly with the spine so that the shock of impact is carried down the length of the spine without twisting or rotating the head at dangerous angles. Of course, neck muscles and tendons are heavy and tough to resist the extreme pressures. The average cow's skull/horn weighs seven pounds; the average bull's weighs 18.

Despite this protection, bulls are not immune to crash damage. Researchers report that warring males often look dazed after clashing. Sometimes they bleed from the nose. One dead bull was found with its skull split open. Another died from a shattered vertebrae. Normally, however, they live to fight another day. After most clashes, each bull bounces back a foot, then steps back several more and shakes its head from side to side like a boxer smiling and waving off a punch as if to say "Is that the best you can do? Never even fazed me." A swinging horn may precede a sudden charge from short range, making it the "wind-up" to increase the impact. Contestants often indulge in considerable pushing and shoving after a clash, much like bison bulls, and try to hook one another in the side, but this tactic rarely succeeds, so quick to anticipate and parry it is each

fighter. Such hooking and in-fighting almost never decides a battle. Although muskox horn tips are potentially deadly, they are reserved almost exclusively for predator defense. In effect, the horns are two weapons in one, the boss being a ram and shield.

Whether they try any pushing or hooking, eventually each bull resumes backing away until, when 30 or more yards separate the two, they rush in for another whack. As if 900 pounds at 25 mph weren't enough, each bull tries to squeeze a little more out of its charge by lunging with its hind legs just before impact, pushing the front legs off the ground, tucking the muzzle down and in, and throwing the horn boss forward and down. Unlike sheep, which turn the head for a one-horn punch, muskoxen land equally with both broad horn helmets. They do not, apparently, seek to gain an advantage by charging from an uphill position, and who could blame them? Additional velocity could only increase the pain.

Sometimes as few as four clashes will settle an altercation, the loser turning his head and walking or running away. Matched bulls have backed and clashed as many as 20 times in a row. Eventually, though, someone always loses and is chased from the field a short distance, horned in the rump if possible. The loser plods purposefully away, never even looking back. The victor celebrates with a bout of sod tearing and roaring. He'll

Muskoxen are built for high speed collisions. Beneath four inches of horn helmet lies a pneumatic skull more than three inches thick. The first vertebrae fits into a notch at the back of the skull, aligning it perfectly with the spine to parry the shock of impact.

walk to each chunk of sod torn up and gland-marked by his antagonist during the threat display and mark it with his own scent, essentially rubbing out every vestige of his vanquished foe. Then he'll indulge in an aggressive round of courting.

Compared to most herbivores, muskoxen have a long rut, beginning as early as July, when dominant bulls move in with cow/calf herds and push other bulls out. Breeding doesn't usually begin until the first week in September when the vast majority of cows reach estrus and are serviced. Then the bulls hang around well into October to pick up the few late bloomers. Why the long courtship? There are at least three probable reasons. One, the presence and odor of the bull may be a catalyst to stimulate cows into a coordinated heat, much as wallowing bull moose do. Two, by grabbing a harem early, a bull increases his chances of hitting the gene pool jackpot. It is easier to defend a harem than take one, and by restricting its movements, a herd bull can reduce the chances that other bulls will discover it. Three, the massive male may need the time to convince his potential breeding partners that he can be trusted. More than most ruminants, bull muskoxen need all the cooperation they can get from cows to ensure a successful mating. Bulls are not designed for sexual gymnastics.

Getting to know one another is the main goal of courtship. If a buck or bull could simply jump in from the hinterlands and do his job the day his services were required, he probably would. But nature doesn't work that way. Most individuals resist close contact with others of their kind (except their own young), so a

bull's toughest assignment is to convince a cow that she should allow him to not only touch her but heave his 900-pound bulk atop her. So Mr. Slow and Easy does not charge in like the bull he is. Though not as cautious as a male mountain goat (muskox cows aren't nearly as nasty and deadly as nannies), he nevertheless acclimates the gals gradually, one by one, beginning with a stare. In most ruminant societies, merely staring at a conspecific (one of your own kind) is not only rude but threatening. So the cow grows nervous. She runs to the safety of the group. But the bull keeps staring, day after day. And she gets used to it. Then he walks toward her, face first. A head-on approach among muskoxen is even more threatening than a stare. After all, that's how clashes start. Again the cow spooks, but she doesn't flee like a bighorn ewe or a whitetail doe. She is afraid to leave the safety of the group, and he isn't exactly chasing her, only walking. So instead of running away, she often runs toward the bull and stands beside him to remove the frightening frontal image of his face and horns. See no evil, fear no evil. Now the bull can indulge in his ultimate deception. He mimics a calf. Much as a human male shaves his beard to appear more childlike and less threatening to women, so a muskox bull stands with his head near a cow's rump, as a calf would to nurse. When she moves, he moves with her, as her calf would. He presses against her. She grows accustomed to his presence and even begins to rub his face with hers.

As estrus draws near, the bull increases his olfactory testing, sniffing each cow's anogenital region, lip curling, and sampling urine. If wolves disturb the herd, then pass on, the excitement spurs him to a frenzy of courting, much as aggressive chasing enflames a bighorn ram's libido. As a cow's estrus draws near, he begins to lay his chin on her rump and back, and kicks her with a front leg, preparing her for the shock of mounting. He is so busy now that half his time is consumed with rut activity. He loses weight rapidly. He begins rushing the cows from behind, bumping into them, halfway climbing them, testing to see if they will stand, for stand they must or he will not succeed in impregnating them at the crucial time.

His persistence and dedication eventually pay off. By the time a cow reaches estrus, she is as steady as a mountain, a mountain he must somehow ascend, and it isn't easy. His chest is so deep and his legs so short that he must pull himself into position by "running" his fore legs along her flanks continuously. He will mount and fail time after time. Researchers have witnessed a single bull try to mount a single cow 26 times in 32 minutes and succeed only five times. How many of those resulted in actual copulation was not known. On average, a bull remained mounted for 10 seconds, five to 10 times longer than required for the nimble bighorn sheep to accomplish the same task.

Somehow the bull manages, for each year calves are born to replenish the herds.

Although the muskox rut seems dull in comparison to the free-for-all of a sheep or caribou rut, energy expenditure is high. That is one of the reasons bulls minimize herding efforts. Keeping large harems together could exhaust a dominant bull before he ever had a chance to breed, increasing the odds that a subdominant could challenge and defeat him. A smart bull, then, husbands his strength for real needs like fighting off challengers and climbing those difficult cows. Females benefit genetically from this strategy by producing male calves with the genetics for large size and fighting power. If they grow to one day dominate a harem and sire many calves, the genes of their progenitors multiply, and that's the whole idea.

MUSKOX RUT: FACTS ON FILE

Family: Bovidae
Genus: Ovibos
Species: moschatus
Weight: Males 700 to 900 lbs., females 500 to 600 lbs.
Height: Males 5 feet, females 3 feet. Horns: 25 to 30 inches long, 30 inches tip to tip, boss 10 inches wide.
Rut: July through October.
Habitat: Lush tundra vegetation, willows, sea coasts.
Range: Arctic Canada and Alaska, Greenland.
Society: Small mixed herds in all seasons. Bachelor bulls alone or in small herds during rut.